The Alfred Russel Wallace Reader

Center Books in Natural History

Shannon Davies
Consulting Editor

George F. Thompson
Series Founder and Director

Published in cooperation with the
Center for American Places,
Santa Fe, New Mexico
and Harrisonburg, Virginia

THE
Alfred Russel Wallace
READER

A Selection of Writings
from the Field

❧ ❦

EDITED BY
Jane R. Camerini

The Johns Hopkins University Press
Baltimore and London

© 2002 The Johns Hopkins University Press
All rights reserved. Published 2002
Printed in the United States of America on acid-free paper
2 4 6 8 9 7 5 3 1

The Johns Hopkins University Press
2715 North Charles Street
Baltimore, Maryland 21218-4363
www.press.jhu.edu

Library of Congress Cataloging-in-Publication Data
Wallace, Alfred Russel, 1823–1913.
[Selections. 2002]
The Alfred Russel Wallace reader : a selection of writings
from the field / edited by Jane R. Camerini.
p.   cm. — (Center books in natural history)
"Books by Alfred Russel Wallace" : p.
Includes bibliographical references (p.   ).
ISBN 0-8018-6781-9 (hardcover : alk. paper) —
ISBN 0-8018-6789-4 (pbk. : alk. paper)
1. Natural selection.   2. Biogeography.
3. Wallace, Alfred Russel, 1823–1913.
I. Camerini, Jane R., 1953–   II. Title.   III. Series.
QH375 .W332 2002
576.8′2—dc21   2001000539

A catalog record for this book is available from the British Library.

Frontispiece: Wallace at home at Old Orchard in Broadstone.
The original photograph from the private collection of Quentin
Keynes is inscribed "From the grandson of Alfred Russel Wallace
to the great grandson of Charles Darwin, Greetings!
Sept. 25, '84, A. J. R. Wallace."

to ugo

~ Contents ~

⁓ *Illustrations* ⟋

~ *Foreword* ~

Who was Alfred Russel Wallace? A complicated man, and so the answer is complicated too. In 1837, he was an unassuming fourteen-year-old English lad from a down-at-heels family, leaving school to go into apprenticeship as a surveyor. In 1907, he was an eighty-four-year-old pundit well known to the British reading public, much honored by scientific societies but also to some degree marginalized as a crank, whose latest of many books was titled *Is Mars Habitable?* Variously, in the years between, he was a leading proselytizer of the Land Nationalization movement (1881), an outspoken opponent of vaccination (1890), and an engaged participant at the International Congress of Spiritualists (1898). But none of these aspects of his life accounts for the book you now hold in your hands. What does account for and justify Jane Camerini's deftly shaped *Reader* is that, amid all of the rest, during the years of his young and middle manhood, Wallace was a field biologist. He spent twelve years in the tropics—first in the Amazon, later in what was then called the Malay Archipelago—collecting insects and other specimens, observing patterns, taking note of small facts while pondering large questions. He was self-educated and self-trained, but he did the work extraordinarily well.

That and more. Although superlatives are as imprudent in judgments of science as in judgments of art, a case can be made that Alfred Russel Wallace was the greatest field biologist of the nineteenth century. If not him, then who? Well, the obvious alternate candidate would be Charles Darwin, but this brings us to a finer point of categorical definition—and I did say *field* biologist. Darwin was indisputably the greatest *conceptual* biologist of his era, yes, and maybe the greatest of all eras. Darwin was also a painstaking experimentalist, a brilliant collator of available data, a masterful architect and cogent articulator of the new biological paradigm. But, notwithstanding his five years aboard the *Beagle*, circling the world as a young gentleman guest of the British navy, exploring and collecting when shore leave allowed, Charles Darwin was never the same sort of adventuresome, fearless, independent, stalwart, and nearly tireless biological traveler that Alfred Russel Wallace was. After the *Beagle* journey, which Darwin both loved and loathed, and which he found painfully too long, he spent the rest of his life as a fretful, cerebral homebody. Wallace, by contrast, spent four years under arduous conditions in the Amazon, survived a ship's sinking on the way home, then promptly headed out again for the East Indies. He had questions still to answer. He had work still to do. This time it was eight years before he returned.

Wallace traveled rough. He paid dues: years of loneliness, precarious sea crossings in small native boats, cold rice, leaky huts, infected feet, malaria. And yet besides enduring the hardships of a consummate field man, Wallace himself also emerged as Darwin's closest peer as a conceptual biologist. Most famously, he co-originated the idea of evolution by natural selection. Less famously but just as notable, he wrote one of the founding texts of what we now call biogeography, *The Geographical Distribution of Animals.* From the boy surveyor had grown a man who saw the profound significance of how landscapes are divided by natural barriers and who recognized the importance of studying which species live where.

Of course, there was another reason besides intellectual hunger for Wallace staying in the field as long as he did. Unlike Darwin, who inherited affluence, Wallace was obliged to earn a living. He did that, first, by shipping natural history specimens to an agent in London for sale to dilettantish collectors; later, he did it from his writings. Although he applied for a few steady jobs after resettling in England, he never landed one, and throughout his long life he remained a relatively impecunious freelancer, fitting his skills and his interests to what the market would reward—a salubrious dilemma he shared with Shakespeare and Leonardo da Vinci, if not with Darwin. Luckily for Wallace, and luckily for us, at his best he had the same aptitude for crafting vivid prose as he had for stalking butterflies in a jungle.

His best was what came from the field. This important distinction is astutely reflected in Jane Camerini's set of well-chosen and well-framed samples. She has spared us the anti-vaccination polemics and the lengthy defenses of spiritualism, even while suggesting their importance within the larger fabric of Wallace's thought. She has skipped Mars, to which

Wallace never went. Her subtitle itself clues what I have just belabored: A Selection of Writings *from the Field.* That is where Alfred Russel Wallace showed his greatest skills and percipience, and that is where, that is why, we should remember him.

David Quammen

~~ *Acknowledgments* ~~

I would like to extend my gratitude to Mark Barrow, Janet Browne, and Martin Fichman, who read the manuscript and provided supportive suggestions. Warm thanks to Richard Russel Wallace, a grandson of Alfred Russel, for his hospitality and generous help over the years. Charles Smith helped to track down Wallace articles for me, and Eliana Camargo provided translations of Portuguese and native Indian words. Closer to home, I would like to thank Louise Robbins for reading and editing an early draft, and Elizabeth Uhr, who gave me many constructive comments.

The illustrations are mostly from Wallace's own books, with the exception of the frontispiece and two maps designed to help readers identify places mentioned in the text. I would like to thank Kim Sholly, of the Darkroom, who worked patiently with me to prepare all of the photographs from Wallace's works, and the University of Wisconsin Cartographic Laboratory staff, who made the maps of the Amazon basin and place names of the Malay Archipelago. The frontispiece photograph of Wallace, published in a book for the first time here, belongs to Quentin Keynes, the great-grandson of Charles Darwin. I am very grateful to Mr. Keynes for his permission to reproduce this candid picture of Wallace, and to Richard Milner, who kindly made this photograph available to me. I thank the Royal Geographical Society of London for permission to publish a portion of Wallace's manuscript map of the Rio Negro, which (to my knowledge) has not been published previously, and the Linnean Society of London for permission to reproduce two original drawings from its collection of Wallace materials. The Departments of Medical Photography and of Special Collections at the University of Wisconsin, Madison, kindly made the jacket illustration available for reproduction.

Carol Zimmerman and Mahinder Kingra at the Johns Hopkins University Press have been most supportive and helpful in seeing this project through to completion. Finally, I wish to thank Randall B. Jones and George F. Thompson at the Center for American Places, whose assiduous efforts made this book possible.

~๏ *A Note to the Reader* ๏~

The purpose of this book is to acquaint readers with the life and work of Alfred Russel Wallace. Known mostly for having spurred Darwin into getting his theory into print, Wallace wrote on a vast range of subjects in addition to evolution, including anthropology, spiritualism, land nationalization, anti-vaccination, and physical geography. The readings in this volume reflect these wide-ranging interests, but I omitted selections on anti-vaccination and spiritualism, available elsewhere, in order to include writings that span Wallace's lifetime. I have chosen whole chapters or articles rather than excerpts from his writings, except for selections from his autobiography and the portion of his 1869 book review in Chapter Four.

Although I hope that my scholarly friends find this book of interest, I wrote it for general readers, not for historians, for whom Wallace may be a familiar figure. To this end, I have kept endnotes and references to a minimum. In order to keep current and nineteenth-century sources separated, the notes at the end of the book have full citations for books or correspondence contemporary with Wallace and are not repeated in the bibliography. Current sources are referenced by name and date in the notes, and are fully cited in the bibliography. All editorial clarifications of Wallace's writings are enclosed in square brackets.

I have listed the first editions of Wallace's books, but many of his books were published later in slightly revised editions. Where I refer to specific page numbers for Wallace's autobiography, *My Life*, or his *Malay Archipelago*, I am using the first American editions of these works, for which details are given in the list of his books.

~~ The Alfred Russel Wallace Reader ~~

Introduction

Biographical Sketch

October 6, 1858

My dear Mother, . . . I have just returned from a short trip and am now about to start on a longer one, but to a place where there are some soldiers, a doctor and engineer who speak English, so if it is good for collecting I shall stay there some months. It is Batchian, an island on the southwest side of Gilolo, about three or four days' sail from Ternate. I am now quite recovered from my New Guinea voyage and am in good health.

I have received letters from Mr. Darwin and Dr. Hooker, two of the most eminent naturalists in England, which has highly gratified me. I sent Mr. Darwin an essay on a subject on which he is now writing a great work. He showed it to Dr. Hooker and Sir C. Lyell, who thought so highly of it that they immediately read it before the Linnean Society. This assures me the acquaintance and assistance of these eminent men on my return home.[1]

The thirty-five-year-old naturalist Alfred Russel Wallace had reason to reassure his mother, having survived hunger, malarial fevers, ulcerated sores, and seasickness in the preceding year. But he was far from ready to return home to England. On the contrary, he felt firmly committed to collecting animal specimens for his research on the origin and distribution of species. Wallace had just learned that his now-famous article on evolution had been published as part of a joint contribution with Charles Darwin in the *Proceedings of the Linnean Society*, a solid accomplishment for an aspiring philosophical naturalist. His confidence was buoyed further by having recently returned from an arduous and successful trip to the Aru Islands near New Guinea, where he had collected the much sought-after great bird of paradise. Thus, halfway through his stay in the Malay Archipelago, we find Wallace vigorously defending his wish to remain in the region to continue his work. In other letters from this period, Wallace expressed his love of solitude and his disinclination for financial and business affairs. These traits endured. Wallace never held a title or a salaried position, and he continued to do research and to write, following his own convictions while remaining an astute observer of natural and human events.

Who was this strong-willed philosophical naturalist? Although Wallace's best-known claim to fame is as co-discoverer, along with Charles Darwin, of the theory of evolution by natural selection, Wallace's interests ranged so broadly that it is difficult to apply a single label, even that of a naturalist, to him. Describing him as a natural scientist would do for the early part of his life, but so would geographer and travel writer; one would have to add social critic, spiritualist, and intellectual for the second half. His status within the scientific community is equally hard to pin down. Historians have called him an outsider, a loner, or the "other" man who discovered evolution, but these terms reflect the slant of historians more than they describe Wallace. Part of the reason he is difficult to categorize is that his concerns were so encompassing and wide ranging. Wallace wrote for the lay person as well as the specialist, and he wrote about biology, evolution, education, religion, morality, spiritualism, vaccination, eugenics, social values, and political systems.

Many of the issues and controversies that engaged him are as vital today as they were 150 years ago. One striking example is his warning about the fate of tropical plant and animal species, which he predicted would "perish irrecoverably" if governments and scientific institutions did not take immediate steps to make complete collections. Whether one looks at major themes of his work—the significance of evolution in moral, political, and social thought, the importance of land and how it should be managed—or at specific elements of his thought, his disagreements with Darwin, or his support of women's rights, one might well be struck by his wisdom, by his ability to focus on things that matter. Yet his views antagonized many people; his infusion of spiritual and theological beliefs into his evolutionary theories, his hearty opposition to vaccination, his support of socialism as he grew older, all made him open to criticism both then and now and have probably contributed to a certain neglect by historians and scientists.

He was an idiosyncratic and enigmatic figure whose long and rich life—he was ninety when he died—is difficult to comprehend in a single glance. Those curious enough to puzzle over this unusual figure have a remarkable quantity of writing available to peruse. In addition to his nine hundred-page autobiography, he published some twenty-one books and seven hundred articles, essays, and letters in newspapers and periodicals. It is not necessary to read the whole of his literary legacy to appreciate his intelligence, perceptiveness, and clear writing. The voluminous published record reveals little of his personal life, about which he was always rather private.[2] The excerpts from his writings reproduced here provide a partial portrayal of this remarkable man.

His writing reveals an independence of thought, which, though shaped by his time and place, bears the stamp of his thoughtfulness, curiosity, and open-mindedness, along with his penchant for being opinionated and outspoken. Quirky, plain-spoken, and modest, he takes us with him as he hikes the Welsh countryside, blazes Amazonian trails, finds his ways through fourteen thousand miles of islands and seas in Indonesia, and then settles into domestic stability as well as scientific respectability, punctuated by travels in Britain, Switzerland, and North America. Through his adventures and honors, his failures and disappointments, Wallace maintained a lack of pretension, an optimism, and enormous compassion, which come through in his writing and capture the heart of many a reader.

Born January 8, 1823, Wallace grew up in a rather poor family, what we might call lower middle class, in rural Wales and then in Hertford, England. During the second quarter of the nineteenth century the economic base of Britain (England, Scotland, and Wales) changed from agriculture to manufacturing, and the British Empire grew in size and power.

For many British people at this time, particularly the working and middle classes, it was a period of growing dissent and new opportunities. The traditional values of Britain's agricultural society were based on wealth and the status of one's family. Industrialization and expansion into colonial lands destabilized the staid social structure, and a substantial redistribution of wealth and power resulted from the need for expertise and scientific knowledge in the transition to a manufacturing society. This broad context shaped Wallace's youth, a context charged with challenges to traditional forms of authority, specifically the Church of England and political power based on birthrights.

Wallace was the eighth of nine children, and his family could barely afford the six years of formal education he received at the one-room Hertford Grammar School. His father, Thomas Vere Wallace, was trained as a lawyer but did not practice law, preferring a quiet country life of gardening and literary pursuits. He lost his modest inheritance in two risky investments; as his family grew in size they moved several times to locations where rent and food were more affordable. Wallace's mother, Mary Anne Greenell, was from an old Hertford family. Wallace tells us almost nothing about her in his autobiography. Of the nine children she bore, only three lived past early adulthood: John, Fanny, and Alfred.

Although Wallace's education was cut short by the family's worsening financial situation, his home was a rich source of books, maps, and gardening activities which Wallace later recalled with pleasure. In his autobiography, he devoted far more time to the games he played as a child than to the lessons he received at school, which he found painful and boring. But he must have been a reasonably good student because in his last year there he assisted in teaching the younger students. This anomalous position of being both pupil and teacher was particularly disagreeable to the tall young man, and for twenty years he suffered from recurrent dreams of great anguish at school. Although Wallace portrayed himself as shy and later extolled the virtues of solitude and reflection, he had many lasting relationships. One, his neighbor and childhood friend, George Silk, remained close to him throughout his life.

Wallace's parents belonged to the Church of England, and as a child Wallace attended services, but he stopped going as soon as he was on his own, at the age of fourteen. His lack of enthusiasm for organized religion became more pronounced when he was exposed to the secular teachings at a Mechanics' Institute, the London Hall of Science.[3] Living in London with his brother John, an apprentice carpenter, the fourteen-year-old Wallace became familiar with the lives of tradesmen and laborers, and he shared in their efforts at self-education. Here Wallace read treatises and attended lectures by Robert Owen and his son which formed the basis of his religious skepticism and his reformist and socialist political philosophy.[4]

Wallace became an apprentice in the surveying firm of his eldest brother, William, in 1837. New tax laws, the division of public lands, and the laying of railway tracks each required surveyors. For approximately eight of the next ten years Wallace surveyed and mapped in Bedfordshire and then in Wales. He spent a great deal of time outdoors, both for work and for pleasure, developing a lifelong interest in land forms, plants, and animals. He became acquainted with many Welsh farmers, but rather than drinking with them in his free time, he preferred to study and write. Among his earliest writings is the essay reproduced here, "The South-Wales Farmer," which shows Wallace's talent for recording his detailed observations even as a very young man. In 1846, his brother William died, and Wallace worked in London and Wales with his brother John, trying to settle their brother's accounts, surveying for a proposed railway line, and designing a building for a Mechanics' Institute at Neath. An enthusiastic amateur naturalist with an intellectual bent, he read widely in the natural history literature of the day, including works by William Swainson, Charles Darwin, and Alexander von Humboldt.[5] He also studied phrenology (a study based on the idea that character and intellect were functions of the brain and could be investigated by the size and shape of the human skull) and mesmerism (a set of techniques whereby one person affects another's mind or body, typically producing a trance), which were controversial and popular pursuits in Britain in the 1840s. Wallace's strong interest in mental phenomena grew throughout his life.

In 1844, unable to find work as a surveyor, Wallace taught at the Collegiate School, a boys' school in Leicester. Here he met Henry Walter Bates, with whom he shared a growing interest in natural history. Like many readers of the sensational bestseller *Vestiges of the Natural History of Creation*,[6] Wallace was inspired by the idea of transmutation, as evolution was then called. Unlike the other tens of thousands of readers of *Vestiges*, he believed that he could find evidence for evolution through the detailed study of one group of animals. Deeply intrigued by the question of the origin of species, unemployed, and ardent in his love of nature, in 1847 Wallace proposed that he and Bates travel to South America as self-employed specimen collectors. This was, as Wallace himself put it, a rather wild scheme. Imagine two young men, loaded with collecting nets and boxes, guns they had recently bought and learned to use, camping equipment, various instruments —including lenses, compasses, thermometers, barometers, and a sextant —and practically no money, taking off for the Amazon to support themselves by selling exotic insects to British collectors of natural history specimens. Although it was clearly a risky adventure, the two naturalists had not only armed themselves with collecting gear, but they also had the practical skills needed for collecting, the knowledge base for identifying what to collect, and the critical link through an agent to the expanding number of natural history collectors.

Although natural objects had been collected and displayed by the wealthy and by noblemen for centuries, their value was enhanced by the growth of a larger set of cultural practices in early Victorian Britain in the 1830s and 40s. The study of nature had multiple social anchors: it was unquestionably fashionable, valued as morally uplifting, encouraged by the Anglican belief in natural theology (in which the study of nature reveals God's greatness), actively pursued scientifically and intellectually, and, although lacking in clear government support, it was sanctioned by cultural values. Natural history collections at the British Museum and anatomical and paleontological collections at the College of Physicians and Surgeons burgeoned while government commissions floundered over how the collections should be run. Professional societies grew in number and membership, providing amateurs such as Wallace opportunities not open at the more elitist, older societies such at the Royal and Linnean Societies. Changes in printing technology, the development of lithography and steam-driven printing presses, along with tax reforms that reduced the high price of paper and postage, meant that natural history books and magazines were readily available. Manufacturers and merchants responded to these changes in Britain, as well as to expanding markets abroad, with a plethora of spectacles and objects, ranging from street shows and private museum displays to affordable illustrated book series, from instruction manuals to microscopes, vials, tools, and outfits of various kinds.[7]

By the time Wallace and Bates left for Brazil in 1848, private collections of exotic natural curiosities were more popular than ever, as was travel literature with a natural history bent; London housed the auction halls and shops where a worldwide trade in natural history specimens was conducted. Wallace's agent, Samuel Stevens, was a professional specimen dealer and enthusiastic collector whose brother, J. C. Stevens, owned a well-known auction house that specialized in zoological specimens. J. C. Stevens's auction house ran regular advertisements in publications such as the *Zoologist* and the *Annals and Magazine of Natural History*. Samuel Stevens not only advertised Bates's and Wallace's wares directly, he also published extracts of their letters to him in these natural history journals. The letters included detailed accounts of Wallace's and Bates's collecting adventures and provide glimpses of both the excitement and the difficulties of their trade.[8]

The two amateur naturalists made their way from Pará (now Belém), Brazil, to inland ports along the Amazon River, using letters of introduction to local merchants and landowners who would in turn assist them in many ways—finding lodgings, providing men to help carry their supplies, and supplying local transportation (often leaky canoes). Although collecting insects and birds was their main task, the practical realities of getting around consumed a great deal of their time. Exigencies of weather, unreliable help, local bureaucratic rules for passports (even within Brazil) and for customs, along with debilitating insect bites and fevers, hindered their efforts to reach good sites. Wallace's younger brother, Herbert Edward, joined them in July 1849 but proved to be more of a poet than an eager assistant. On his way back to England in 1851 Herbert Edward Wallace died of yellow fever, which was raging through the port of Pará, where he waited for passage home.

After collecting together for nearly two years, Bates and Wallace separated in March 1850, Wallace choosing to explore the Rio Negro, and Bates the Upper Amazon. Bates struggled with sickness and loneliness, yet continued collecting in Brazil for eleven years,[9] whereas Wallace spent a total of four years traveling, collecting, mapping, drawing, and writing in unexplored regions of the Amazon basin (see Figure 5). He studied the languages and habits of the peoples he encountered, collected insects (especially butterflies), birds, fish, and mammals, all the while hoping to find clues to the mystery of the origin of plant and animal species. Between long excursions on the Rio Negro, Wallace visited Richard Spruce, an English botanist who had come to Brazil in 1849, and together they examined their collections and talked about species. In Britain and America, the idea of common descent was profoundly controversial, but in the Amazon, Wallace was unperturbed by the conflicts surrounding the publication of *Vestiges*. The writings reproduced here show Wallace broaching the subject indirectly because he knew he had not yet found the clear evidence he sought.

When Wallace headed for home, his baggage consisted of numerous cases and boxes of bird skins, insects, stuffed alligators, drawings, and some thirty-four live animals. In addition, six boxes he had packed the previous year had never been sent, so he had to pay customs charges for these as well as for the rest of his cargo. Encumbered by his crates and live animals, still weak with fever, Wallace boarded the brig Helen and sailed for England on July 12, 1852. The bulk of his collections were lost when his ship went up in flames and sank. His on-the-scene description of his disastrous journey is included here in Chapter Two.

Wallace managed to save some of his notes and drawings, and from these and from memory he published six scientific articles, two books, *Palm Trees of the Amazon and Their Uses* and *A Narrative of Travels on the Amazon and Rio Negro* (both 1853), and a map depicting the course of the Rio Negro.[10] Although the map won him favorable notice from the Royal Geographical Society, which helped to fund his next collecting venture, his reception in London was mixed. He presented papers at the Zoological and Entomological Societies as a visitor, not a member, despite the relative openness of these societies to men with expertise. Wallace was criticized by a few members as a "mere collector" who had no business theorizing. Lacking university training, family money or support, professional connections, and a proper accent, Wallace carried on, undaunted by his critics and determined to keep searching.

The selections of his work presented here include an excerpt from his Amazonian narrative and portions of both the manuscript and published versions of his map of the Rio Negro. His paper on the monkeys of the Amazon shows how clear his scientific writing was and how forthright he was in his insistence on the importance of precise observations and records of geographical distribution. Among the products of his trip not reproduced here are some of his scientific achievements: he named and described what were, at the time, fourteen new species of palm and fourteen "type specimens" of birds (specimens on which the formal name and description of the spe-cies are based), and he brought home a set of drawings of Amazonian fish.

He not only learned the ropes of tropical collecting from his journey, but his curiosity about species origins was keener than ever. Wallace made his next trip, to the East Indian (Malay) Archipelago, after studying the collections at the British Museum and sifting through advice from his agent, the curators at the British Museum, and other professional contacts in London. He believed that the animals from the region would be of commercial value and scientific interest and that the practical conditions for collecting would be manageable because he could take advantage of Dutch and English settlements there. Fifteen months after returning from Brazil, Wallace departed on a steamship, traveled overland from Alexandria to Suez, and sailed to Singapore, where he arrived in April 1854.

Wallace spent eight years in the Malay Archipelago (1854–62), traveling widely among the islands, collecting biological specimens for his own research and for sale, and writing scores of scientific articles, mostly on zoological subjects. He would be well known to naturalists for his collections alone, amassing as he did more than 125,000 specimens, hundreds of which were new to science. His collections were so successful that he had a modest income to live on when he returned to England, which would have provided some lasting financial security had he let Stevens continue to manage his money. As it turned out, Wallace made poor investment decisions and lost his capital; in subsequent years he sold most of his own collections because he needed money, and he got by on book sales, fees for grading examinations, and a small government pension.

Wallace's field work in Indonesia and New Guinea in the mid-nineteenth century required perseverence and skill, but it also required the help of local people. The fine class distinctions that underlay the social order in Britain were blurred in the East, where Wallace was received as a gentleman and a scientist. European society in the East was the creation of European colonialism, especially the Dutch settlements in Java, Borneo, and the Moluccas, and the English trade routes between India and China, routes that had evolved during two thousand years of spice-trading practices, long before Europeans arrived. Wallace used the roads, houses, and other buildings that enabled colonial agriculturalists, miners, and engineers to tap the resources of these islands. The doctors, missionaries, traders, teachers, and merchants who built their lives within the new mixed cultures served Wallace as critical points of departure for his collecting. It was through their hospitality that he found lodgings, boats or horses for traveling, directions to places for collecting, servants, translators, introductions to local rulers, and letters of introduction to other Europeans. Native servants were also crucial to his success, as was Stevens, who sent him information and supplies regularly. Yet Wallace's years in the Malay were far more than a collector's success story; it was here that he proved himself a philosophical naturalist of the highest order.

Among the scientific papers he wrote in the Malay are two extraordinary articles dealing with the origin of new species. The first of these, "On the Law Which Has Regulated the Introduction of New Species" (referred to as the "Law" essay), published in 1855, asserted a necessary connection between the geological and geographical existence of species. Just as one species replaces a closely related form in geographical space, so too a species in the fossil record is most closely related to one in the nearest fossil layer. The reason for this necessary connection in space and in time is common descent—if species come from pre-existing species, it follows that the most likely place to find the ancestor of a given form is underneath it in the fossil record or nearby in geographical terms. After carefully arguing a series of propositions consistent with known patterns of geographical and geological distribution, Wallace concluded that "every species has come into existence coincident both in space and time with a pre-existing closely allied species." This remarkable argument combined geological history with the growing mass of information about geographical distribution to make a compelling hypothesis about the origin of species. The famous geologist Sir Charles Lyell began his own notebook on species after reading Wallace's article. Darwin basically ignored it, much to his later dismay. Bates wrote from the Upper Amazon with compliments to his friend, noting that few people would truly understand it.[11] In his next theoretical essay, the one for which he is now best known, Wallace went on to propose that new species arise by the progression and continued divergence of varieties, which outlive the parent species in the struggle for existence. Wallace sent these ideas early in

1858 to Darwin, who saw such a striking similarity to his own unpublished but extensive work on evolution that he chose to publish some of his previous writings along with Wallace's paper, resulting in a joint publication in the Proceedings of the Linnean Society (for more details, see Chapter Three).

Unlike his theoretical papers, Wallace's publications on the classification and geographical distribution of animals among the islands of the archipelago were welcome contributions that did not stir up priority concerns. His articles on the natural history of the Aru Islands, on the orangutan, and on birds of paradise were read with great interest. His detailed biogeographical analyses led him to formulate what is now known as "Wallace's Line," the boundary that separates the fauna of Australia from that of Asia (see Figure 1). For several generations, zoologists had tried various methods of characterizing the animals unique to each region of the world, and the Asian and Australian faunas had been particularly difficult to delineate. Wallace's local observations of patterns in insect, bird, and mammal distribution enabled him to improve upon existing schemes of zoological regions.[12] For Wallace, the analyses of patterns of distribution were inseparable from his evolutionary theory; he presented his novel findings on distribution without making common descent an issue, but the assumption of it pervades his scientific writings from the Malay.

Following his return to England in 1862, Wallace forged ahead with his work on the animal variation and distribution, climate, geography, dispersal, and evolution. He lived for several years in London, where he frequented numerous scientific societies and enjoyed the company of many learned men, especially Sir Charles Lyell. It was clearly a source of satisfaction and stimulation to the self-taught naturalist to be a welcome participant in the conversations of London's intellectual elite, including Darwin, Thomas H. Huxley, Herbert Spencer, John Stuart Mill, and his old friends Henry Bates and Richard Spruce, among others. But Wallace did not belong to any particular bandwagon. Never one to bow to the opinion of others, regardless of their stature, Wallace became an outspoken advocate for many social, political, scientific, civic, and medical causes. He participated actively in making science accessible to a wide audience, but his role as a public intellectual was more than as a commentator on scientific controversy; his was also a reflective political, social, philosophical, and spiritual commentary. His open-mindedness (or

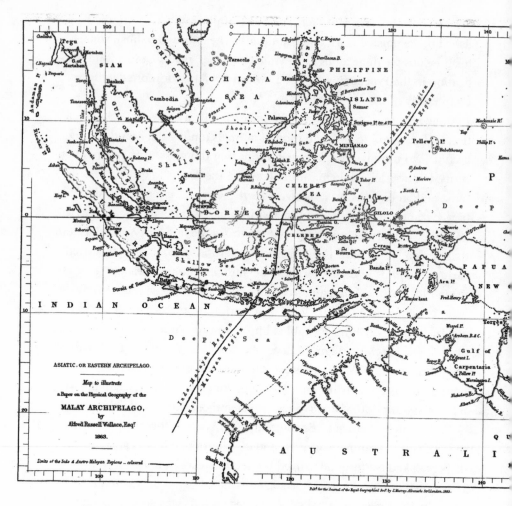

Fig. 1. Map of Wallace's Line. In Wallace's day it was well known that the boundary between the characteristic animals of the Australian and of the Asian regions meet in the Malay Archipelago, but precisely where to locate this boundary captivated naturalists for more than two centuries. Wallace synthesized known and new observations of the geography, geology, and zoology of the region in several letters and articles before creating this map of his results. The *solid diagonal line* marks the division between the Asian (Indo-) and Australian faunas; the *dashed line* east of it indicates the boundary separating the Malayan from the Papuan peoples. From "On the Physical Geography of the Malay Archipelago," *Journal of the Royal Geographical Society of London,* 33 (1863):217–34; facing p. 217.

credulity, as some biographers and some of his friends called it), however, led him to embrace some speculative theories and questionable ap-
proaches that have not stood the test of time. But as a public intellectual,
he could be counted on to hold forth with compassion and reason on a
wide range of subjects, and his opinions were taken seriously if not always accepted.

Urban life and scientific societies, although allowing Wallace to continue his scientific work as well as pursue his other interests, did not hold
him for very long. After a drawn-out engagement to a lady who rebuffed
him, Wallace married Annie Mitten, daughter of the eminent botanist
William Mitten, in 1866, and they had three children (Herbert, who died
at age six, Violet, and William). The Wallace family moved several times
(beginning in 1870 they moved from inner London to its eastern out-
skirts at Barking, then further east to Grays in Essex, then south to Dorking, Croydon, Godalming, and Parkstone) before finally settling at Broadstone in Dorset. Wallace built three of the houses, and at each home he
and Annie, a knowledgeable botanist in her own right, made a new garden. Wallace's delight in gardening only increased as he got older, and he
cultivated orchids, grew hundreds of other uncommon plants, and built
greenhouses, ponds, and aquatic tanks. Although the Wallace home
changed locations many times for various reasons, it was a stable and
happy one. The warmth of his family life is apparent in the reminiscences of his children (see Chapter Four).

Wallace's intellectual life extended far beyond science; he sought to in-
corporate his growing involvement in spiritual matters, as well as his social and political views, into a larger vision. It was a purposeful universe
that Wallace struggled to articulate, one in which he drew together the
science of organic evolution with his commitments to socialism and
spiritualism. In the grand scheme that emerges gradually through his
writings over half a century, Wallace included not only questions about
human society, but plants as well as animals, fossils as well as living organisms, and even questions about the possibility of life on other planets
and the distribution of stars. As he aged and his perspective grew more
encompassing, he increasingly saw in nature the workings of an intelligent mind authoring nature for a purpose: the spiritual development of
man. His later views contradicted earlier ones when, for example, he had
interpreted the incredible beauty of birds of paradise as proof that all

living things were *not* made for man (in Chapter Three, see Wallace's reaction to finding his first king bird of paradise).

In the second half of the nineteenth century, the writings of Karl Marx had a strong following on the European continent and elsewhere, while in North America and Britain alternative conceptions of socialism held sway. Earlier in his life Wallace held a mixture of ideals about the well-being of society which had elements of two distinct ideologies. On the one hand, some of his opinions were allied closely to Herbert Spencer's philosophy of the evolution of society based on strong individualism. On the other hand, Wallace leaned closer to socialism in his conviction that private land ownership led to social and economic injustice and in his appreciation of Robert Owen's experiments with communities based on cooperation and education for every individual. In response to an article he published in 1880 on land nationalization, a group of people formally organized the Land Nationalization Society. Wallace served as its first president, and over the years he devoted a good deal of time to studying and writing proposals for a more just system of land ownership, most notably his book *Land Nationalisation* (1882). By 1889 Wallace had become firmly convinced that he had to abandon the strong individualism of the philosophies of Herbert Spencer and John Stuart Mill for the promise of socialism, which he had come to believe was the only social system that could provide equality and justice to every human being. Drawing heavily on two books by the American socialist Edward Bellamy,[13] Wallace applied socialist philosophy to his scientific view of human selection. He argued that only under socialism could natural selection play its part in improving the physical, mental, and moral qualities of humans. Wallace felt this to be his single most important and original contribution, that is, the idea that the only way for natural selection to continue to improve the human species was "under a social system which gives equal opportunities of culture, training, leisure, and happiness to every individual."[14]

In the meantime, in 1869 Wallace had published a highly successful narrative of his journey, *The Malay Archipelago: The Land of the Orang-utan and the Bird of Paradise*. The most famous of all books on the Malay Archipelago, Wallace's narrative was by far his most favorably received work (and Joseph Conrad's favorite book). Wallace employed the talents of some of the finest London engravers to prepare wood engravings for the illustrations, which were based on his own sketches, on photographs, and on actual specimens. *The Malay Archipelago* went through ten editions before 1900, was translated into German and Dutch, and excerpts were published in magazines and in the *Library of the World's Best Literature*.[15]

Wallace saw something more than natural selection at work in the evolutionary process, as he made clear in a review published in 1869, part of which is the first selection in Chapter Four. In 1870 his *Contributions to the Theory of Natural Selection* appeared. In the last two essays of this work and in several articles on human evolution and on spiritualism, Wallace parted from the scientific naturalism of many of his colleagues in claiming that natural selection could not account for the higher faculties, the moral, artistic, and mathematical capacities, of human beings (see the second selection in Chapter Four). Although his views about human evolution continued to diverge from those of Darwin and Huxley, among others, Wallace was hardly alone in his attempt to discover natural laws that made room for a supreme intelligence. Many scientists, theologians, and philosophers sought their own version of a theistic approach to evolution, with the theism allied to Christianity. But Wallace, scornful of organized religion, carved out a central role for a non-Christian spiritual evolution.

Although these views isolated him from some of his scientific colleagues, his willingness to take on the so-called supernatural as a subject worthy of serious scientific investigation connected him to many other people. Not only did he reach wide audiences with his numerous essays and books on spiritualism, he made evolution palatable to many who found Darwin too technical or too heretical. And he continued to produce some of his most important and comprehensive scientific works, publishing his two-volume *The Geographical Distribution of Animals* in 1876 and *Island Life* in 1880. Even though some of the scientific ideas he supported, such as the permanence of continents and oceans and the astronomical causes of glaciers, have not endured, his books were received enthusiastically and became classics in zoogeography and island biogeography. They were the first full-length works that synthesized knowledge of the distribution and dispersal of living and extinct organisms in an evolutionary framework. As a collector, travel and science writer, and evolutionist, Wallace's has left us a valuable legacy. But his efforts to combine the insights of biology with more complex human concerns are still not adequately understood.

Wallace's status as the neglected "other man" who formulated the theory of evolution by natural selection is a twentieth-century phenomenon, the work of historians who saw Wallace chiefly as a spur to Darwin, or those who saw Wallace as having been undeservedly obscured for various reasons. Wallace did not lack recognition during his lifetime. The list of honors, awards, and honorary degrees he received is impressive indeed, and shortly after his death he was memorialized at Westminster Abbey.[16] It was in the following half-century that his role was minimized. Clearly, Darwin was the hero of choice for scientists and historians building the modern evolutionary synthesis and raising the status of biology as a science. During the last few decades of the twentieth century several biographies were published, as well as several books exploring the relationship of Darwin and Wallace. The questions surrounding the Darwin-Wallace "case," as one writer calls it, remain largely unresolved.

Wallace's concern with spiritual and mental phenomena is the most conspicuous of the topics that call for a further study; much of the published work to date is unduly dismissive of Wallace's participation in mesmerism and related subjects. The science of mesmerism and other controversial areas were pivotal to changes in medical and scientific authority in the nineteenth century.[17] Wallace's concerns with land nationalization and social reform have also not yet been given careful attention. Further analysis of his unpublished correspondence and journals is likely to reveal a more nuanced picture of Wallace's character; there are hints that he was irritable with editors, righteous in his own way with colleagues, and on occasion misleading, even mendacious.[18] He was neither a loner nor an outsider when one looks closely at his relationships with George Silk, Stevens, Ali, Spruce, Bates, other colleagues, and family members. As I write, more than one new biography is in preparation, and our understanding of Wallace as a whole will be enhanced by these efforts to capture the complexity of his work and his character in a more finely tuned historical context. (Raby 2001 has just been published.)

In the meantime, this selection of Wallace's writings will acquaint readers with one of the most engaging thinkers and writers of the Victorian Age. Unquestionably an astute observer of the natural world, of landscapes, and of human social conditions, he also saw himself through the eyes of those he observed. He recorded the following reaction of a native in Malaysia: "One day when I was rambling in the forest, an old man stopped to look at me catching an insect. He stood very quiet till I had pinned and put it away in my collecting box, when he could contain himself no longer, but bent almost double, and enjoyed a hearty roar of laughter."[19]

1

Wales

Wallace was fourteen years old when he set out to learn the trade of land surveying under his eldest brother, William, who had a small business. As the readings here show, he took great pleasure in mastering the skills of surveying, in living in the countryside of England and Wales, and in getting to know the farmers, laborers, and landlords with whom he lived. He absorbed firsthand the political and social circumstances in which they lived. The significance of the land boundaries he was helping to lay down and document was not apparent to him at the time, but in later years, the relationship of workers to the land and the question of who should own the land would occupy a central role in his thought. The 1830s and 40s were years of great agitation in Britain, with dissidents from the middle and working classes demanding a range of economic and political reforms, such as free trade and a limit to governmental control of prices, universal male suffrage, and more equable representation for the working classes. With revolution a threat throughout Europe, the British aristocracy responded slowly with modest reforms to the needs of farmers and factory workers in the rapidly industrializing towns and cities. As a surveyor, Wallace measured and drew boundaries for new tax laws, for the apportionment of public land to existing landowners, and for new lines for the laying of railroads.

During the decade from 1837 to 1847, Wallace transformed himself from a timid youth into an itinerant natural scientist (Figure 2). His autobiography, written many years after the events he records, is replete with descriptions of friends he made, the homes and habits of local farmers and artisans, and the newspapers and books he borrowed from them. During a lull in surveying when the brothers could not find work, Wallace spent a year and a half as a schoolteacher. As mentioned in the Introduction, his friendship with Henry Walter Bates enhanced his increasingly serious study of natural history. At the same time a burgeoning interest

Fig. 2. Alfred R. Wallace in 1848, at age twenty-five, before leaving for Brazil. From *My Life,* 1: facing p. 266.

in mental phenomena took shape while he was a teacher at Leicester. When he began surveying he knew virtually nothing about geology, plants, or animals. Ten years later, self-taught in the rudiments of natural history, he had resolved to travel to the tropics to try to solve the basic question in natural history: how do species originate? Some of the experiences that led to this seemingly radical transition are related here in selections from Wallace's autobiography. But the readings document more than the change from surveyor to naturalist; they provide a glimpse of the political and social education that laid the ground for his passionate commitment to reform.

In the first excerpt, the brothers are hired to survey a parish "for the commutation of the tithes" in a county north of London. The tithe, which had been previously paid in produce, was now a rent or tax that farmers had to pay in cash on the value of their produce and on the quality of the land. A new law, the Tithe Commutation Act, had been enacted the preceding year, 1836, which required that accurate surveys be drawn to calculate the productivity of the land. Landowners wanted to exact the rent due to them, and tenant farmers were required to pay higher taxes on their produce. The Wallace brothers made detailed maps of each field keyed to lists of landowners, occupiers, type of cultivation, square area, and tithe rent charge.[1] Tenant farmers resented the cash demands made on them, and by 1842, their resentment turned violent in some quarters.

Similarly, when the brothers were hired to make surveys for the enclosure of common land, Wallace thought that although it was a pity to enclose a wild place, there must be a sound reason behind the law requiring it. Over the course of several decades he came to hold strong and radical opinions against enclosure and for the nationalization of land. The brief second selection from his autobiography is taken from the period when the brothers were paid to prepare maps for the enclosure of public lands in South Wales. Wallace explains that the General Enclosure Act was supposed to turn unproductive land into cultivated, productive land. But what he saw later was that by taking this common, public land, which poor farmers used for pasture, and giving it to landowners, the wealthy were made wealthier, and the poor farmers lost the right to use the land. Here, he expresses his outrage at the unfairness of the enclosures; his sense of social justice in relation to the land is powerfully and eloquently condensed in this short passage.

During the weeks and months when he and his brother could find no surveying work, Wallace spent a lot of time alone, reading, walking, and following the interests that his outdoor life had stimulated. The next selection is a retrospective account of how the world of botany opened up to him. Discovering the order and diversity in the plant kingdom was a delight, but learning to identify the plants he found was a pivotal experience; love and curiosity for learning about nature began to preoccupy him and would soon take him far afield from the Welsh countryside for which he felt a growing fondness. Included in his autobiography is a photograph of one particularly picturesque waterfall in the vale of Neath near which Wallace walked, collecting plants and insects (Figure 3).

The fourth selection records Wallace's observations from the years he spent surveying in Wales, about six years altogether. We can see that he is already a rather opinionated young man, critical of many Welsh habits and harsh in his words for the established church. His observational

Fig. 3. Ysgwd Gladys, one of the most interesting cascades around which Wallace walked in the vale of Neath. From *My Life,* 1: facing p. 248.

Fig. 4. Free Library of Neath, a building designed by Alfred R. and John Wallace for the Mechanics' Institute. From *My Life*, 1: facing p. 246.

skills come through, resulting in a detailed, textured picture of rural life. In the introduction to this sketch, "The South-Wales Farmer," Wallace mentions that interest in Wales had been piqued by the recent Rebecca disturbances, a series of violent actions taken by small groups of agitators in response to new tolls that farmers were charged to use roads. The tolls became a focal point for dissatisfaction among tenant farmers and workingmen, for whom the new road tax was added to their growing list of grievances. The increase in tithe rents, land enclosures, food shortages, and increased prices had all contributed to worsening conditions for the poor, and some used the gates of the new toll houses as targets for their protests.[2]

Wallace's years of surveying would soon draw to a close. His brother William died unexpectedly in early 1846, and in the last selection of this chapter we find Wallace unhappily trying to tie up his brother's accounts and responding to wild swings in the surveying market because of speculation and competition for the building of new railroad lines. Even though he was poor, Wallace resided in relative stability in the small city of Neath, where he participated in the local workingman's institute as an instructor and builder (Figure 4). It was here that he set his sights on his future work; just after the events described in the reading, his friend Henry Bates came to visit. They spent time collecting beetles and making plans for a collecting trip to the tropics. By autumn 1847 the young naturalist's goal had crystallized, as described in a letter to Bates, "I begin to feel rather dissatisfied with a mere local collection. I should like to take some one family to study thoroughly, principally with a view to the theory of the origin of species. By that means I am strongly of opinion that some definite results might be arrived at."[3]

❧ ❧
Wallace Begins Surveying

It was, I think, early in the summer of 1837 that I went with my brother William into Bedfordshire [a county about thirty miles northwest of Lon-

My Life, 1:105–10.

don] to begin my education as a land-surveyor. The first work we had was to survey the parish of Higham Gobion for the commutation of the tithes. It was a small parish of about a thousand acres, with the church, vicarage, and a good farmhouse on the highest ground, and a few labourer's cottages scattered about, but nothing that could be called a village. The whole parish was one large farm; the land was almost all arable and the fields very large, so that it was a simple piece of work. We took up our quarters at the Coach and Horses public-house in the village of Barton-in-the-Clay, six miles north of Luton, on the coach-road to Bedford. We were nearly a mile from the nearest part of the parish, but it was the most convenient place we could get.

An intelligent young labourer was hired to draw the chain in measuring, while I carried a flag or measuring-rod and stuck in pegs or cut triangular holes in the grass where required, to form marks for future reference. We carried billhooks for cutting rods and pegs, as well as for clearing away branches that obstructed the view, and for cutting gaps in the hedges on the main lines of the survey, in order to lay them out perfectly straight. We started work after an early breakfast, and usually took with us a good supply of bread-and-cheese and half a gallon of beer, and about one o'clock sat down under the shelter of a hedge to enjoy our lunch. My brother was a great smoker, and always had his pipe after lunch (and often before breakfast), and, of course, the chain-bearer smoked too. It therefore occurred to me that I might as well learn the art, and for a few days tried a few whiffs. Then, going a little too far, I had such a violent attack of headache and vomiting that I was cured once and for ever from any desire to smoke, and although I afterwards lived for some years among Portuguese and Dutch, almost all of whom are smokers, I never felt any inclination to try again.

Three miles north of Barton was the small village of Silsoe adjoining Wrest Park, the seat of Earl Cowper, whose agent, Mr. Brown, was known to my brother, and had, I think, obtained from him the parish survey we were engaged upon. A young gentleman three or four years older than myself who was, I think, a pupil of Mr. Brown's, was sent by him to learn a little land-surveying with us, and was a pleasant companion for me, especially as we were often left alone, when my brother was called away on other business, sometimes for a week at a time. Although the country north of Barton was rather flat and uninteresting, to the south it was very picturesque, as it was only about half a mile from the range of the North Downs, which, though only rising about three hundred feet above Barton, yet were very irregular, jutting out into fine promontories or rounded knolls with very steep sides and with valleys running up between them. The most charming of these valleys was the nearest to us, opening behind the church. It was narrow, with abundance of grass and bushes on the sides of a rapid-flowing streamlet, which, about a quarter of a mile further, had its source in a copious spring gushing out from the foot of the chalk-hill. On the west side of this valley the steep slope was thickly covered with hazel and other bushes, as well as a good many trees, forming a hanging wood full of wild flowers, and offering a delightful shade in the heat of the afternoon. About a mile to the east there was an extensive old British earthwork called Ravensburgh Castle, beyond which was another wooded valley; between these was a tolerably level piece of upland where the villagers played cricket in the summer.

My friend, whose name I forget (we will call him Mr. A.), was a small-sized but active young fellow, very good-looking, and quite the dandy in his dress. He was proud of his attractions, and made friends with any of the good-looking village girls who would talk to him. One day we met a pretty rosy-cheeked girl about his own age—a small farmer's daughter—and after a few words, seeing she was not disinclined for a chat, he walked back with her, and I went home. When he returned, he boasted openly of having got her to promise to meet him again, but the landlord advised him to be careful not to let her father see him. A day or two after, as we were passing near the place, he saw the girl again, and I walked slowly on. I soon heard loud voices, and, looking back, saw the girl's father, a big, formidable-looking man, threatening the young Lothario with his stick, and shouting out that if he caught him there again with *his* girl, he would break every bone in his body. When the young gentleman came back he was not the least abashed, but told us the whole story very much as it had happened, and rather glorying in his boldness in not running away from so big and enraged a man, and intimating that he had assuaged his anger by civil words, and had come away with flying colours.

One day he and I went for a walk over the hills towards Hitchin, where on the ordnance map a small stream was named Roaring Meg, and we wanted to see why it was so called. We found a very steep and narrow valley something like that called the Devil's Dyke near Brighton; but this was thickly wooded on both sides, and the little stream at the bottom, rushing over a pebbly bed, produced a roaring sound which could be heard at a considerable distance. This northern range of downs has the advantage over the south downs of having numerous springs and streams on both sides of it, and these are especially abundant around the ancient village of Tod-dington, five miles west of Barton, where the ordnance map shows about twenty springs, the sources of small streams, within a radius of two miles.

It was while living at Barton that I obtained my first information that there was such a science as geology, and that chalk was not *everywhere* found under the surface, as I had hitherto supposed. My brother, like most land surveyors, was something of a geologist, and he showed me the fossil oysters of the genus Gryphæa and the Belemnites, which we had hitherto called "thunderbolts," and several other fossils which were abundant in the chalk and gravel around Barton. While here I acquired the rudiments of surveying and mapping, as well as calculating areas on the map by the rules of trigonometry. This I found very interesting work, and it was rendered more so by a large volume belonging to my brother, giving an account of the great Trigonometrical Survey of England, with all the angles and the calculated lengths of the sides of the triangles formed by the different sta-tions on hilltops, and by the various church spires and other conspicuous objects. [The survey began in 1784 as a means of determining the locations of the Paris and Greenwich Observatories; it evolved into the Ordnance Survey of Great Britain, producing detailed topographic maps of the entire country.] The church spires of Barton and Higham Gobion had been thus used, and the distance between them accurately given; and as the line from one to the other ran diagonally across the middle of the parish we were sur-veying, this was made our chief base-line, and the distance as measured found to agree very closely with that given in the survey. This volume was eagerly read by me, as it gave an account of all the instruments used, in-cluding the great theodolite, three feet in diameter, for measuring the an-gles of the larger triangles formed by distant mountain tops, often twenty or thirty miles apart, and in a few cases more than a hundred miles; the accurate measurement of the base-lines by steel chains laid in wooden troughs, and carefully tightened by exactly the same weight passing over a pulley, while the ends were adjusted by means of microscopes; the exact temperature being also taken by several thermometers in order to allow for contraction or expansion of the chains; and by all these refinements sev-eral base-lines of seven or eight miles in length were measured with ex-treme accuracy in distant parts of the country. These base-lines were tested by repeated measurements in opposite directions, which were found to differ only by about an inch, so that the mean of all the measurements was probably correct to less than half that amount.

These bases were connected by the system of triangulation already referred to, the angles at all the stations being taken with the best available instruments and often repeated by different observers, while allowance had also to be made for height above the sea-level, to which all the distances had to be reduced. In this way, starting from any one base, the lengths of the sides of all the triangles were calculated, and ultimately the length of the other bases; and if there had been absolutely no error in any of the measurements of base-lines or of angles, the length of a base obtained by calculation would be the same as that by direct measurement. The results obtained showed a quite marvellous accuracy. Starting from the base measured on Salisbury Plain, the length of another base on the shore of Lough Foyle in the north of Ireland was calculated through the whole series of triangles connecting them, and this calculated length was found to differ from the measured length by only five inches and a fraction. The distance between these two base-lines is about three hundred and sixty miles.

These wonderfully accurate measurements and calculations impressed me greatly, and with my practical work at surveying and learning the use of that beautiful little instrument, the pocket-sextant, opened my mind to the uses and practical applications of mathematics, of which at school I had been taught nothing whatever, although I had learnt some Euclid and algebra. This glimmer of light made me want to know more, and I obtained some of the cheap elementary books published by the Society for the Diffusion of Useful Knowledge [a publishing organization formed to make scientific, literary, and artistic knowledge accessible to the working classes through numerous inexpensive publications. It was founded in 1826 and folded in 1846.] The first I got were on Mechanics and on Optics, and for some years I puzzled over these by myself, trying such simple experiments as I could, and gradually arriving at clear conceptions of the chief laws of elementary mechanics and of optical instruments. I thus laid the foundation for that interest in physical science and acquaintance with its general principles which have remained with me throughout my life.

🖝 It was here, too, that during my solitary rambles I first began to feel the influence of nature and to wish to know more of the various flowers, shrubs, and trees I daily met with, but of which for the most part I did not even know the English names. At that time I hardly realized that there was such a science as systematic botany, that every flower and every meanest and most insignificant weed had been accurately described and classified, and that there was any kind of system or order in the endless variety of plants and animals which I knew existed. This wish to know the names of wild plants, to be able even to speak of them, and to learn anything that was known about them, had arisen from a chance remark I had overheard about a year before. A lady, who was governess in a Quaker family we knew at Hertford, was talking to some friends in the street when I and my father met them, and stayed a few minutes to greet them. I then heard the lady say, "We found quite a rarity the other day—the Monotropa; it had not been found here before." This I pondered over, and wondered what the Monotropa was. All my father could tell me was that it was a rare plant; and I thought how nice it must be to know the names of rare plants when you found them. However, as I did not even know there were books that described every British plant, and as my brother appeared to take no interest in native plants or animals, except as fossils, nothing came of this desire for knowledge till a few years later. . . .

~∙≻ ≺∙~

Enclosing Common Land

When we had finished at Llanbister, we went about ten miles south to a piece of work that was new to me—the making of a survey and plans for the enclosure of common lands. This was at Llandrindod Wells [a town in southeastern Wales], where there was then a large extent of moor and mountain surrounded by scattered cottages with their gardens and small fields, which, with their common rights, enabled the occupants to keep a horse, cow, or a few sheep, and thus make a living. All this was now to be taken away from them, and the whole of this open land divided among the landowners of the parish or manor in proportion to the size or value of their estates. To those that had much, much was to be given, while from the poor their rights were taken away; for though nominally those that *owned* a little land had some compensation, it was so small as to be of no use to them

My Life, 1:149–50.

in comparison with the grazing rights they before possessed. In the case of all cottagers who were tenants or leaseholders, it was simple robbery, as they had no compensation whatever, and were left wholly dependent on farmers for employment. And this was all done—as similar enclosures are almost always done—under false pretences. The "General Enclosure Act" states in its preamble, "Whereas it is expedient to facilitate the enclosure and improvement of commons and other lands now subject to the rights of property which obstruct cultivation and the productive employment of labour, be it enacted," etc. But in hundreds of cases, when the commons, heaths, and mountains have been partitioned out among the landowners, the land remains as little cultivated as before. It is either thrown into adjacent farms as rough pasture at a nominal rent, or is used for game-coverts, and often continues in this waste and unproductive state for half a century or more, till any portions of it are required for railroads, or for building upon, when a price equal to that of the best land in the district is often demanded and obtained. I know of thousands of acres in many parts of the south of England to which these remarks will apply, and if this is not obtaining land under false pretences—a legalized robbery of the poor for the aggrandizement of the rich, who were the law-makers—words have no meaning. . . .

❧ ☙
Wallace Discovers Botany in Wales

During the larger portion of my residence at Neath we had very little to do, and my brother was often away, either seeking employment or engaged upon small matters of business in various parts of the country. I was thus left a good deal to my own devices, and having no friends of my own age I occupied myself with various pursuits in which I had begun to take an interest. Having learnt the use of the sextant in surveying, and my brother having a book on Nautical Astronomy, I practised a few of the simpler observations. Among these were determining the meridian by equal altitudes of the sun, and also by the pole-star at its upper or lower culmination; finding the latitude by the meridian altitude of the sun, or of some of the principal stars; and making a rude sundial by erecting a gnomon towards the

My Life, 1:190–95.

pole. For these simple calculations I had Hannay and Dietrichsen's Almanac, a copious publication which gave all the important data in the Nautical Almanac, besides much other interesting matter, useful for the astronomical amateur or the ordinary navigator. I also tried to make a telescope by purchasing a lens of about two feet focus at an optician's in Swansea, fixing it in a paper tube and using the eye-piece of a small opera glass. With it I was able to observe the moon and Jupiter's satellites, and some of the larger star-clusters; but, of course, very imperfectly. Yet it served to increase my interest in astronomy, and to induce me to study with some care the various methods of construction of the more important astronomical instruments; and it also led me throughout my life to be deeply interested in the grand onward march of astronomical discovery.

But what occupied me chiefly and became more and more the solace and delight of my lonely rambles among the moors and mountains, was my first introduction to the variety, the beauty, and the mystery of nature as manifested in the vegetable kingdom.

I have already mentioned the chance remark which gave me the wish to know something about wild flowers, but nothing came of it till 1841, when I heard of and obtained a shilling paper-covered book published by the Society for the Diffusion of Useful Knowledge, the title of which I forget, but which contained an outline of the structure of plants and a short description of their various parts and organs; and also a good description of about a dozen of the most common of the natural orders of British plants. Among these were the Cruciferæ, Caryophylleæ, Leguminosæ, Rosaceæ, Umbelliferæ, Compositæ, Scrophularineæ, Labiatæ, Orchideæ, and Glumaceæ. This little book was a revelation to me, and for a year was my constant companion. On Sundays I would stroll in the fields and woods, learning the various parts and organs of any flowers I could gather, and then trying how many of them belonged to any of the orders described in my book. Great was my delight when I found that I could identify a Crucifer, an Umbellifer, and a Labiate; and as one after another the different orders were recognized, I began to realize for the first time the order that underlies all the variety of nature. When my brother was away and there was no work to do, I would spend the greater part of the day wandering over the hills or by the streams gathering flowers, and either determining their position from my book, or coming to the conclusion that they belonged to other orders of which I knew nothing, and as time went on I found that there were a very large number of these, including many of our most beautiful and curious flowers, and I felt that I *must* get some other book by which I could learn something about these also. But I knew of no suitable book, I did not even know that any British floras existed, and having no one to help me I was obliged to look among the advertisements of scientific or educational publications that came in my way. At length, soon after we came to Neath, David Rees happened to bring in an old number of the *Gardener's Chronicle*, which I read with much interest, and as I found it in advertisements and reviews of books, I asked him to bring some more copies, which he did, and I found in one of them a notice of the fourth edition of Lindley's "Elements of Botany," which, as it was said to contain descriptions of all the natural orders, illustrated by numerous excellent woodcuts, I thought would be just the thing to help me on. The price, 10s. 6d., rather frightened me, as I was always very short of cash; but happening to have so much in my possession, and feeling that I *must* have some book to go on with, I ordered it at Mr. Hayward's shop.

When at length it arrived, I opened it with great expectations, which were, however, largely disappointed, for although the larger part of the book was devoted to systematical botany, and all the natural orders were well and clearly described, yet there was hardly any reference to British plants—not a single genus was described, it was not even stated which orders contained any British species and which were wholly foreign, nor was any indication given of their general distribution or whether they comprised numerous or few genera or species. The inclusion of all the natural orders and the excellent woodcuts illustrating many of them, and showing the systematic characters by dissections of the flowers and fruits, were, however, very useful, and enabled me at once to classify a number of plants which had hitherto puzzled me. Still, it was most unsatisfactory not to be able to learn the names of any of the plants I was observing, so one day I asked Mr. Hayward if he knew of any book that would help me. To my great delight he said he had Loudon's "Encyclopædia of Plants," which contained all the British plants, and he would lend it to me, and I could copy the characters of the British species.

I therefore took it home to Bryn-coch, and for some weeks spent all my leisure time in first examining it carefully, finding that I could make out both the genus and the species of many plants by the very condensed but clear descriptions, and I therefore copied out the characters of every British species there given. As Lindley's volume had rather broad margins, I found room for all the orders which contained only a moderate number of species, and copied the larger orders on sheets of thin paper, which I interleaved at the proper places. Having at length completed this work for all the flowering plants and ferns, and also the genera of mosses and the main divisions of the lichens and fungi, I took back the volume of Loudon, and set to work with increased ardour to make out all the species of plants I could find. This was very interesting and quite a new experience for me, and though in some cases I could not decide to which of two or three species my plant belonged, yet a considerable number could be determined without any doubt whatever.

This also gave me a general interest in plants, and a catalogue published by a great nurseryman in Bristol, which David Rees [a farmer with whom Wallace lodged for a year] got from the gardener, was eagerly read, especially when I found it contained a number of tropical orchids of whose wonderful variety and beauty I had obtained some idea from the woodcuts in Loudon's Encyclopædia. The first epiphytal orchid I ever saw was at a flower show in Swansea, where Mr. J. Dillwyn Llewellyn exhibited a plant of *Epidendrum fragrans*, one of the less attractive kinds, but which yet caused in me a thrill of enjoyment which no other plant in the show produced. My interest in this wonderful order of plants was further enchanced by reading in the *Gardener's Chronicle* an article by Dr. Lindley on one of the London flower shows, where there was a good display of orchids, in which, after enumerating a number of the species, he added, "and Dendrobium Devonianum, too delicate and beautiful for a flower of earth." This and other references to and descriptions of them gave them, in my mind, a weird and mysterious charm, which was extended even to our native species, and which, I believe, had its share in producing that longing for the tropics which a few years later was satisfied in the equatorial forests of the Amazon.

But I soon found that by merely identifying the plants I found in my walks I lost much time in gathering the same species several times, and even then not being always quite sure that I had found the same plant before. I therefore began to form a herbarium, collecting good specimens and drying them carefully between drying papers and a couple of boards weighted with books or stones. My brother, however, did not approve of my devotion to this study, even though I had absolutely nothing else to do, nor did he suggest any way in which I could employ my leisure more profitably.

He said very little to me on the subject beyond a casual remark, but a letter from my mother showed me that he thought I was wasting my time. Neither he nor I could foresee that it would have any effect on my future life, and I myself only looked upon it as an intensely interesting occupation for time that would be otherwise wasted. Even when we were busy I had Sundays perfectly free, and used them to take long walks over the mountains with my collecting box, which I brought home full of treasures. I first named the species as nearly as I could do so, and then laid them out to be pressed and dried. At such times I experienced the joy which every discovery of a new form of life gives to the lover of nature, almost equal to those raptures which I afterwards felt at every capture of new butterflies on the Amazon, or at the constant stream of new species of birds, beetles, and butterflies in Borneo, the Moluccas, and the Aru Islands.

It must be remembered that my ignorance of plants at this time was extreme. I knew the wild rose, bramble, hawthorn, buttercup, poppy, daisy, and foxglove, and a very few others equally common and popular, and this was all. I knew nothing whatever as to genera and species, nor of the large numbers of distinct forms related to each other and grouped into natural orders. My delight, therefore, was great when I was now able to identify the charming little eyebright, the strange-looking cow-wheat and louse-wort, the handsome mullein and the pretty creeping toad-flax, and to find that all of them, as well as the lordly foxglove, formed parts of one great natural order, and that under all their superficial diversity of form there was a similarity of structure which, when once clearly understood, enabled me to locate each fresh species with greater ease. The Crucifers, the Pea tribe, the Umbelliferæ, the Compositæ, and the Labiates offered great difficulties, and it was only after repeated efforts that I was able to name with certainty a few of the species, after which each additional discovery became a little less difficult, though the time I gave to the study before I left England was not sufficient for me to acquaint myself with more than a moderate proportion of the names of the species I collected.

✍ Now, I have some reason to believe that this was the turning-point of my life, the tide that carried me on, not to fortune, but to whatever reputation I have acquired, and which has certainly been to me a never-failing source of much health of body and supreme mental enjoyment. If my brother had had constant work for me so that I never had an idle day, and if I had continued to be similarly employed after I became of age, I should most probably have become entirely absorbed in my profession, which, in its various departments, I always found extremely interesting, and should therefore not have felt the need of any other occupation or study. . . .

~❧ ❧~

The South-Wales Farmer
Introductory Remarks

In the following pages I have endeavoured to give a correct idea of the habits, manners, and mode of life of the Welsh hill farmer, a class which, on account of the late Rebecca disturbances, has excited much interest. Having spent some years in Radnorshire, Brecknockshire, Glamorganshire, and other parts of South Wales, and been frequently in the dwellings of the farmers and country people, and had many opportunities of observing their customs and manners, all that I here mention is from my own observation, or obtained by conversation with the parties. I have taken Glamorganshire as the locality of most of what I describe, as I am best acquainted with that part and the borders of Carmarthenshire, where the recent disturbances have been most prominent.

Whenever there is any great difference in neighbouring counties, I have noticed it. I may here observe that in Radnorshire the Welsh manners are in a great measure lost with the language, which is entirely English, spoken with more purity than in many parts of England, with the exception of those parts bordering Cardiganshire and Brecknockshire, where the Welsh is still used among the old people, the River Wye, which is the boundary of the latter county and Radnorshire, in its course between Rhayader Gwy and the Hay, also separates the two languages. On the Radnorshire side of the river you will find in nine houses out of ten English commonly spoken, while directly you have crossed the river, there is as great or a still greater preponderance of Welsh. In the country a few miles round the seaport town of Swansea most of the peculiarities I shall mention may be seen to advantage. In the east and south-eastern parts of Glamorganshire, called the Vale of Glamorgan, the appearance of the country and the inhabitants is much more like those of England. The land is very good and fertile, agri-

My Life, 1:205–22.

culture is much attended to and practised on much better principles. This part, therefore (the neighbourhood of the towns of Cowbridge and Cardiff), is excepted from the following remarks.

The South-Wales Farmer: His Modes of Agriculture, Domestic Life, Customs, and Character

The generality of mountain farms in Glamorganshire and most other parts of South Wales are small, though they may appear large when the number of acres only is considered, a large proportion being frequently rough mountain land. On the average they consist of from twenty to fifty acres of arable land in fields of from four to six, and rarely so much as ten acres; the same quantity of rough, boggy, bushy, rushy pasture, and perhaps as much, or twice as much, short-hay meadow, which term will be explained hereafter; and from fifty to five hundred acres of rough mountain pasture, on which sheep and cattle are turned to pick up their living as they can.

Their system of farming is as poor as the land they cultivate. In it we see all the results of carelessness, prejudice, and complete ignorance. We see the principle of doing as well as those who went before them, and no better, in full operation; the good old system which teaches us not to suppose ourselves capable of improving on the wisdom of our forefathers, and which has made the early polished nations of the East so inferior in every respect to us, whose reclamation from barbarism is ephemeral compared with their long period of almost stationary civilization. The Welshman, when you recommend any improvement in his operations, will tell you, like the Chinaman, that it is an "old custom," and that what it did for his forefathers is good enough for him. But let us see if the farmer is so bad as this mode of doing his business may be supposed to make him. In his farmyard we find the buildings with broken and gaping doors, and the floors of the roughest pitching. In one corner is a putrid pond, the overflowings of which empty themselves into the brook below. Into this all the drainings from the dungheaps in the upper part of the yard run, and thus, by evaporation in summer and the running into the brook in winter, full one-half of the small quantity of manure he can obtain (from his cattle spending the greater part of their time on the mountain and in wet bushy pastures) is lost.

The management of his arable land is dreadfully wasteful and injurious.

Of green crops (except potatoes can be so called) he has not the slightest idea, and if he takes no more than three grain crops off the land in succession, he thinks he does very well; five being not uncommon. The first and principal crop is wheat, on which he bestows all the manure he can muster, with a good quantity of lime. He thus gets a pretty good crop. The next year he gets a crop of barley without any manure whatever, and after that a crop of oats, unmanured. He then leaves the field fallow till the others have been treated in the same manner, and then returns to serve it thus cruelly again; first, however, getting his potato crop before his wheat. Some, after the third crop (oats), manure the land as well as they can, and sow barley with clover, which they mow and feed off the second year, and then let it remain as pasture for some time; others, again, have three crops of oats in succession after the wheat and barley, and thus render the land utterly useless for many years.

In this manner the best crops of wheat they can get with abundance of manure, on land above the average quality, is about twenty bushels per acre—ten bushels is, however, more general, and sometimes only seven or eight are obtained.

The rough pastures on which the cattle get their living and waste their manure a great part of the time consist chiefly of various species of rushes and sedges, a few coarse grasses, and gorse and fern on the drier parts. They are frequently, too, covered with brambles, dwarf willows, and alders.

The "short-hay meadows," as they are called, are a class of lands entirely unknown in most of England; I shall, therefore, endeavour to describe them.

They consist of large undulating tracts of lands on the lower slopes of the mountains, covered during autumn, winter, and spring with a very short brownish yellow wet turf. In May, June, and July the various plants forming this turf spring up, and at the end of summer are mown, and form "short-hay"; and well it deserves the name, for it is frequently almost impossible to take it up with a hayfork, in which case it is raked up and gathered by armfuls into the cars. The produce varies from two to six hundredweight per acre; four may be about the average, or five acres of land to produce a ton of hay. During the rest of the year it is almost good for nothing. It is astonishing how such stuff can be worth the labour of mowing and making it into hay. An English farmer would certainly not do it, but the poor Welshman has no choice; he must either cut his short-hay or have no food for his cattle in the winter; so he sets to, and sweeps away with his scythe a breadth which would astonish an English mower.

The soil which produces these meadows is a poor yellow clay resting on the rock; on the surface of the clay is a stratum of peaty vegetable matter, sometimes of considerable thickness though more generally only a few inches, which collects and retains the moisture in a most remarkable manner, so that though the ground should have a very steep slope the water seems to saturate and cling to it like a sponge, so much so that after a considerable period of dry weather, when, from the burnt appearance of the surface, you would imagine it to be perfectly free from moisture, if you venture to kneel or lie down upon it you will almost instantly be wetted to the skin.

The plants which compose these barren slopes are a few grasses, among which are the sweet vernal grass (*Anthoxanthum odoratum*) and the crested hair grass (*Kaleria cristata*), several Cyperaceæ—species of carex or sedge which form a large proportion, and the feathery cotton grass (*Eriophorum vaginatum*). The toad-rush (*Juncus bufonius*) is frequently very plentiful, and many other plants of the same kind. Several rare or interesting British plants are here found often in great profusion. The Lancashire asphodel (*Narthecium ossifragum*) often covers acres with its delicate yellow and red blossoms. The spotted orchis (*O. maculata*) is almost universally present. The butterwort (*Pinguicula vulgaris*) is also found here, and the beautiful little pimpernel (*Anagallis tenella*). The louseworts (*Pendicularis sylvatica* and *P. palustris*), the melancholy thistle (*Cincus heterophyllus*), and the beautiful blue milkwort (*Polygala vulgaris*), and many others, are generally exceedingly plentiful, and afford much gratification to the botanist and lover of nature.

The number of sheep kept on these farms is about one to each acre of mountain, where they live the greater part of the year, being only brought down to the pastures in the winter, and again turned on the mountain with their lambs in the spring. One hundred acres of pasture and "short-hay meadow" will support from thirty to forty cattle, ten or a dozen calves and oxen being sold each year.

The farmers are almost invariably yearly tenants, consequently little improvement is made even in parts which could be much bettered by draining. The landlord likes to buy more land with his spare capital (if he has any) rather than improve these miserable farms, and the tenant is too poor to lay out money, or if he has it will not risk his being obliged to leave the farm or pay higher rent in return for his permanently improving another person's land.

The hedges and gates are seldom in sufficiently good repair to keep out cattle, and can hardly be made to keep out mountain sheep, who set them completely at defiance, nothing less than a six-foot stone wall, and not always that, serving to confine them. The farmer consequently spends a good deal of his time in driving them out of his young clover (when he has any) or his wheat. He is also constantly engaged in disputes, and not infrequently litigation, with his neighbours, on account of the mutual trespasses of their stock.

The Welshman is by no means sharp-sighted when his cattle are enjoying themselves in a neighbour's field, especially when the master is from home, otherwise the fear of the "pound" will make him withdraw them after a short time.

On almost every farm water is very plentiful, often far too much so, and it is sometimes run over a meadow, but in such a manner as to lose one-half of the advantage which might be derived from it. The farmer is contented with merely cutting two or three gaps in the watercourse at the top, from which the water flows over the field as it best can, scarcely wetting some parts and making complete pools in others.

Weeding he considers quite an unnecessary refinement, fit only for those who have plenty of money to waste upon their fancies—except now and then, when the weeds have acquired an alarming preponderance over the crop, he perhaps sets feebly to work to extract the more prominent after they have arrived at maturity and the mischief is done. His potatoes are overrun with persicarias, docks, and spurges; his wheat and barley with corn cockle, corn scabious, and knapweed, and his pastures with thistles, elecampine, etc., all in the greatest abundance. If you ask him why he leaves his land in such a disgraceful state, and try to impress upon him how much better crops he would have if he cleared it, he will tell you that he does not think they do much harm, and that if he cleaned them this year, there would be as many as ever next year, and, above all, that he can't afford it, asking you where he is to get money to pay people for doing it.

The poultry, geese, ducks, and fowls are little attended to, being left to pick up their living as well as they can. Geese are fattened by being turned into the corn stubble, the others are generally killed from the yard. The fowls, having no proper places to lay in, are not very profitable with regard to eggs, which have to be hunted for and discovered in all sorts of places. This applies more particularly to Glamorganshire, which is in a great measure supplied with eggs and poultry from Carmarthenshire, or "Sir Gaer" (pronounced there *gar*) as it is called in Welsh, where they manage them much better.

If there happens to be in the neighbourhood anyone who farms on the improved English system, has a proper course of crops, with turnips, etc., folds his sheep, and manages things in a tidy manner, it is impossible to make the Welshman believe that such a way of going on pays; he will persist that the man is losing money by it all the time, and that he only keeps it on because he is ashamed to confess the failure of his new method. Even should the person go on for many years, to all appearance prosperously and in everybody else's eyes making money by his farm, still the Welshman will declare that he has some other source from which he draws to purchase his dear-bought farming amusement, and that the time will come when he will be obliged to give it up; and though you tell him that the greater part of the land in England is farmed in that manner, and ask him whether he thinks they can all be foolish enough to go on losing money year after year, he is still incredulous, says that he knows "the nature of farming," and that such work as that can never pay. While the ignorance which causes this incredulity exists, it is evidently a difficult task to improve him.

Domestic Life, Customs, etc.

The house is a tiled, white-washed edifice, in the crevices of which wall rue, common spleenwort, and yarrow manage generally to vegetate, notwithstanding their (at the very least) annual coat of lime. It consists on the ground floor of a rather large and very dark room, which serves as kitchen and dining-room for the family, and a rather better one used as a parlour on high days or when visitors call; this latter frequently serves as the bedroom of the master and mistress. The kitchen, which is the theatre of the Welsh farmer's domestic life, has either a clay floor or one of very uneven stone paving, and the ceiling is in many cases composed of merely the floor boards of the room above, through the chinks of which everything going on aloft can be very conveniently heard and much seen. The single window is a small and low one, and this is rendered almost useless by the dirtiness of the glass, some window drapery, a Bible, hymn book and some old news-papers on the sill, and a sickly-looking geranium or myrtle, which seems a miracle of vital tenacity in that dark and smoky atmosphere. On one side may be discerned an oak sideboard brilliantly polished, on the upper part of which are rows of willow pattern plates and dishes, in one corner an open cupboard filled with common gaudily-coloured china, and in the other a tall clock with a handsome oak case. Suspended from the ceiling is a serious impediment to upright walking in the shape of a bacon rack, on which is, perhaps, a small supply of that article and some dried beef, also some dried herbs in paper, a large collection of walking sticks, and an old gun. In the chimney opening a coal fire in an iron grate takes the place of the open hearth and smoky peat of Radnorshire and other parts. A long substantial oak table, extending along the room under the window, an old armchair or two, a form or bench and two or three stools, complete the fur-niture of the apartment. From the rack before mentioned is generally sus-pended a piece of rennet for making cheese, and over the mantelpiece is probably a toasting-fork, one brass and two tin candlesticks, and a milk strainer with a hole in the bottom of it; on the dresser, too, will be perceived a brush and comb which serve for the use of the whole family, and which you may apply to your own head (if you feel so inclined) without any fear of giving offence.

Upstairs the furniture is simple enough: two or three plain beds in each room with straw mattresses and home-made blankets, sheets being entirely unknown or despised; a huge oak chest full of oatmeal, dried beef, etc., with perhaps a chest of draws to contain the wardrobe; a small looking-glass which distorts the gazer's face into a mockery of humanity; and a plentiful supply of fleas, are all worth noticing. Though the pigs are not introduced into the family quite so familiarly as in Ireland, the fowls seem to take their place. It is nothing uncommon for them to penetrate even upstairs; for we were once ourselves much puzzled to account for the singular phenome-non of finding an egg upon the bed, which happened twice or we might have thought it put there by accident. It was subsequently explained to us that some persons thought it lucky for the fowls to lay there: the abundance of fleas was no longer a mystery. The bed in the parlour before mentioned serves, besides its ostensible use, as a secret cupboard, where delicacies may be secured from the junior members of the family. I have been in-formed by an acquaintance whose veracity I can rely on (and indeed I should otherwise find no difficulty in believing it) that one day, being asked to take some bread and cheese in a respectable farmhouse, the wheat bread (a luxury) was procured from some mysterious part of the bed, either between the blankets or under the mattress, which my informant could not exactly ascertain. The only assistants in the labours of the farm, besides the sons and daughters, is generally a female servant, whose duties are multifarious and laborious, including driving the horses while ploughing and in haytime, and much other out-of-door work. If you enter the house in the morning, you will probably see a huge brass pan on the fire filled with curdled milk for making cheese. Into this the mistress dips her red and not particularly clean arm up to the elbow, stirring it round most vigorously. Meals seem to be prepared solely for the men, as you seldom see the women sit down to table with them. They will either wait till the others have done or take their dinner on their laps by the fire. The breakfast consists of hasty-pudding or oatmeal porridge, or cheese with thin oatmeal cakes or barley bread, which are plentifully supplied at all meals, and a basin of milk for each person; for dinner there is perhaps the same, with the addition of a huge dish of potatoes, which they frequently break into their basin of milk or eat with their cheese; and for supper, often milk with flummery or "siccan" (pronounced *shiccan*). As this is a peculiar and favourite Welsh dish, I will describe its composition. The oat bran with some of the meal left in it is soaked for several days in water till the acetous fermentation commences; it is then strained off, producing a thin, starchy liquid. When wanted for use this is boiled, and soon becomes nearly of the consistence and texture of blancmange, of a fine light brown colour and a peculiar acid taste which, though at first disagreeable to most persons, becomes quite pleasant with use. This is a dish in high repute with all real Welshmen. Each person is provided with a basin of new milk, cold, and a spoon, and a large dish of hot flummery is set on the table, each person helping himself to as much as he likes (and that is often a great deal), putting it in his basin of milk; and it is, I have no doubt, very wholesome and nourishing food. I must mention that the women, both in the morning and evening (and frequently at dinner too), treat themselves to a cup of tea, which is as universal a necessary among the fair sex here as in other parts of the kingdom. They prefer it, too, without milk, which they say takes away the taste, and as it is generally made very weak, that may be the case. Once or twice a week a piece of bacon or dry beef is added to dinner or supper, more as a relish to get down the potatoes than as being any food in itself. The beef in particular is so very high-dried and hard as almost to defy the carver's most strenuous efforts. The flavour is, nevertheless, at times very fine when the palate gets used to it, though the appearance is far from inviting, being about the colour and not far from the hardness of the black oak table. They generally keep it in a large chest in oatmeal (which was before mentioned). Often, when lodging at a little country inn, have we, when just awake in the morning, seen one of the children come stealthily into the room, open the lid of the huge chest, climb over the edge of it, and, diving down, almost disappear in its recesses, whence, after sundry efforts and strainings, he has reappeared, dragging forth a piece of the aforesaid black beef, which is obtained thus early that it may be soaked a few hours before boiling, to render it most submissive to the knife.

From the foregoing particulars it will be seen that these people live almost entirely on vegetable food. When a cow or a pig is killed, for a day or two they luxuriate on fresh meat; but this is the exception, not the rule. Herrings, too, they are fond of as a relish, as well as cockles and other indigestible food; but neither these nor the beef and bacon can be considered to be the staple food of the peasantry, which is, in one form or another, potatoes, oatmeal, bread, cheese, and milk.

The great consumption of oatmeal produces, as might be expected, cutaneous diseases, though, generally speaking, the people are tolerably healthy. They have a great horror of the doctor, whom they never send for but when they think there is some great danger. So long as the patient is free from pain they think all is right. They have not the slightest idea of what an invalid ought to eat. If gruel is ordered, they make a lumpy oatmeal pudding, to which, however, the sick man will frequently prefer bread and cheese. When they have gone on in this way till the unhappy individual is in the greatest danger and the medical attendant insists upon his directions being attended to, they unwillingly submit; and if the patient dies, they then impute it entirely to the doctor, and vow they will never call him in to kill people again.

As in most rural districts, by constant inter-marriages every family has a host of relations in the surrounding country. All consider it their duty to attend a funeral, and almost every person acquainted with the deceased attends as a mark of respect. Consequently the funerals are very large, often two or three hundred persons, and when the corpse has to be carried a distance, most of them come on horseback, which, with the varied colours of the women's dresses and the solemn sounds of a hymn from a hundred voices, as they wend their way along some lonely mountain road, has a most picturesque and interesting effect. This large company generally meet at the house, where provisions are ready for all who choose to partake of them. The well-known beautiful custom of adorning the graves with flowers and evergreens is much practised.

When a birth takes place in a family all the neighbours and relations call within a few days to inquire after the health of the mother and child, and take a cup of tea or bread and cheese, and everyone brings some present, either a pound of sugar, quarter pound of tea, or a shilling or more in money, as they think best. This is expected to be returned when the givers are in a similar situation.

The "bidding," which is a somewhat similar custom at a marriage, is not quite so general, though it is still much used in Carmathenshire. When a young couple are married they send notice to all their friends, that "on a day named they intend to have a 'bidding,' at which they request their company, with any donations they may think proper, which will be punctually returned when they are called upon on a similar occasion." At such biddings £20 or £30 are frequently collected, and sometimes much more, and as from various causes they are not called upon to return more than one-half, they get half the sum clear, and a loan without interest of the other half to commence life with.

The national dress or costume of the men (if ever they had any) is not now in use; that of the women, however, is still very peculiar. Both use principally home-made articles, spinning their own wool and sending it to the factory to be made into flannel or cloth. They also dye the wool black themselves, using in the operation the contents of certain well-known domestic utensils, which is kept stewing over the fire some days, emitting a most unsavoury odour, which, however, they assert to be very wholesome. The men generally wear a square cut coat of home-made pepper-and-salt coloured cloth, waistcoat and breeches or trousers of the same, and a round low-crowned hat; or occasionally fustian trousers and gay flannel waistcoat with bright metal buttons, coloured neckerchief, home-knit stockings of black sheep's wool, and lace-up boots. Shirts of checked coarse flannel—cotton shirts and sheets being considered equally luxurious. One of the most striking parts of the women's dress is the black beaver hat, which is almost universally worn and is both picturesque and becoming. It is made with a very high crown, narrowing towards the top, and a broad, perfectly flat brim, thus differing entirely from any man's hat. They frequently give thirty shillings for one of these hats, and make them last the greater part of their lives. The body dress consists of what they call a bedgown, or *bet-cown*, as it is pronounced, which is a dress made quite plain, entirely open in front (like a gentleman's dressing gown), with sleeves a little short of the elbow. A necessary accompaniment to this is an apron, which ties it up round the waist. The bedgown is invariably formed of what they call flannel, which is a stuff formed by a mixture of wool, cotton, and sometimes a little silk. It is often striped black or dark blue, or brown and white, with alternate broad and narrow stripes, or red and black, but more frequently a plaid of several colours, the red and black being wool, the white or blue cotton, and often a narrow yellow stripe of silk, made in plaid patterns of every variety of size and colour. The apron is almost always black-and-white plaid, the only variety being in the form and size of the pattern, and has a pretty effect by relieving the gay colours of the other part of the dress. They in general wear no stays, and this, with the constant habit of carrying burdens on the head, produces almost invariably an upright carriage and good figure, though rather inclined to the corpulency of Dutch beauties. On their necks they usually wear a gay silk kerchief or flannel shawl, a neat white cap under the hat; laced boots and black worsted stockings complete their attire. In Carmarthenshire a jacket with sleeves is frequently worn by the women, in other respects their dress does not much differ from what I have described.

The women and girls carry (as before mentioned) great loads upon their heads, fifty or sixty pounds weight, and often much more. Large pitchers (like Grecian urns) of water or milk are often carried for long distances on uneven roads, with both hands full at the same time. They may be often seen turning round their heads to speak to an acquaintance and tripping along with the greatest unconcern, but never upsetting the pitcher. The women are almost invariably stout and healthy looking, notwithstanding their hard work and poor living. These circumstances, however, make them look much older than they really are. The girls are often exceedingly pretty when about fifteen to twenty, but after that, hard work and exposure make their features coarse, so that a girl of five-and-twenty would often be taken for nearer forty.

All, but especially the young ones, ride most fearlessly, and at fairs they may be seen by dozens racing like steeple-chasers.

Many of these farmers are freeholders, cultivating their own land and living on the produce; but they are generally little, if any, better off than the tenants, leaving the land in the same manner, thus showing that it is not altogether want of leases and good landlords that makes them so, but the complete ignorance in which they pass their lives.

All that I have hitherto said refers solely to the poorer class, known as hill farmers. In the valleys and near the town where the land is better, there are frequently better educated farmers, who assimilate more to the English in their agricultural operations, mode of living, and dress.

In all the mining districts, too, there is another class—the colliers and furnacemen, smiths, etc., who are as different from the farmers in everything as one set of men can be from another. When times are good their wages are such as to afford them many luxuries which the poor farmer considers far too extravagant. Instead of living on vegetable diet with cheese and buttermilk, they luxuriate on flesh and fowl, and often on game too, of their own procuring. But in their dress is the greatest difference. The farmer is almost always dressed the same, except that on Sundays and market-day it is newer. But the difference between the collier or furnaceman at his work—when he is half naked, begrimed from head to foot, labouring either in the bowels of the earth or among roaring fires, and looking more like demon than man—and on holidays dressed in a suit of clothes that would not disgrace an English gentleman, is almost remarkable. It is nothing uncommon to see these men dressed in coat and trousers of *fine black cloth*, elegant waistcoat, fine shirt, beaver hat, Wellington boots, and a fine silk handkerchief in his pocket; and instead of being ridiculous, as the clumsy farmer would be in such a dress, wearing it with a quiet, unconcerned, and gentlemanly air. The men at the large works, such as Merthyr Tydfil, are more gaudy in their dress, and betray themselves much more quickly than the colliers of many other districts.

It is an undoubted fact, too, that the persons engaged in the collieries and iron works are far more intellectual than the farmers, and pay more attention to their own and their children's education. Many of them indeed are well informed on most subjects, and in every respect much more highly civilized than the farmer.

The wages which these men get—in good times £2 or £3 per week—prevents them, with moderate care, from being ever in any great distress. They likewise always live well, which the poor farmer does not, and though many of them have a bit of land and all a potato ground, the turnpike grievances, poor-rates, and tithes do not affect them as compared with the farmers, to whom they are a grievous burden, making the scanty living with which they are contented hard to be obtained.

Their rents, too, continue the same as when their produce sold for much more and the above-mentioned taxes were not near so heavy. The consequence is that the poor farmer works from morning to night after his own fashion, lives in a manner which the poorest English labourer would grumble at, and as his reward, perhaps, has his goods and stock sold by his landlord to pay the exorbitant rent, averaging 8s.

Language, Character, etc.

The Welsh farmer is a veritable Welshman. He can speak English but very imperfectly, and has an abhorrence of all Saxon manners and innovations. He is frequently unable to read or write, but can sometimes con over his Welsh Bible, and make out an unintelligible bill; and if in addition he can read a little English and knows the four first rules of arithmetic, he may be considered a well-educated man. The women almost invariably neither read nor write, and can scarcely ever understand two words of English. They fully make up for this, however, by a double share of volubility and animation in the use of their own language, and their shrill clear voices are indications of good health, and are not unpleasant. The choleric disposition usually ascribed to the Welsh is, I think, not quite correct. Words do not often lead to blows, as they take a joke or a satirical expression very good humouredly, and return it very readily. Fighting is much more rarely resorted to than in England, and it is, perhaps, the energy and excitement with which they discuss even common topics of conversation that has given rise to the misconception. They have a ready and peculiar wit, something akin to the Irish, but more frequently expressed so distantly and allegorically as to be unintelligible to one who does not understand their modes of thought and peculiarities of idioms, which latter no less than the former they retain even when they converse in English. They are very proud of their language, on the beauty and expression of which they will sometimes dilate with much animation, concluding with a triumphant assertion that theirs *is* a language, while the English is none, but merely a way of speaking.

The language, though at times guttural, is, when well spoken, both melodious and impressive. There are many changes in the first letters of words, for the sake of euphony, depending on what happens to precede them; *m* and *b*, for instance, are often changed into *f* (pronounced *v*), as *melin* or *felin*, a mill; *mel* or *fel*, honey. The gender is often changed in the same manner, as *bach* (masculine), *fach* (feminine), small; *mawr* (m.), *fawr* (f.), great. The mode of making the plural is to an Englishman rather *singular,* a syllable being taken off instead of being added, as is usually the case with us, as *plentyn*, a child; *plant,* children; *mochyn,* a pig; *moch,* pigs. But in other cases a syllable or letter is added.

Their preachers or public speakers have much influence over them. During a discourse there is the most breathless attention, and at the pauses a universal thrill of approbation. Allegory is their chief specialty, and seems to give the hearers the greatest pleasure, and the language appears well fitted for giving it its full effect.

As might be expected from their ignorance, they are exceedingly superstitious, which is rather increased than diminished in those who are able to read by their confining their studies almost wholly to the Bible. The forms their superstitions take are in general much the same as in Scotland, Ireland, and other remote parts of the kingdom. Witches and wizards and white witches, as they are called, are firmly believed in, and their powers much dreaded. There is a witch within a mile of where I am now writing who, according to report, has performed many wonders. One man who had offended her she witched so that he could not rise from his bed for several years, but was at last cured by inviting the witch to tea and making friends with her. Another case was of a man driving his pig to market when the witch passed by. The pig instantly refused to move, sat up on its hind legs against the hedge in such a manner as no pig was ever seen to do before, and, as it could not be persuaded to walk, was carried home, where it soon died. These and dozens of other similar stories are vouched for by eye-witnesses, one of whom told me this. A still more extraordinary instance of the woman's supernatural powers must be mentioned. She is supposed to have the power of changing herself into different shapes at pleasure, that of a hare seeming to be with her, as with many other witches, the favourite one, as if they delighted in the persecution that harmless animal generally meets with. It is related that one day, being pursued by men and dogs in this shape, the pursuers came to a coal mine the steam engine of which was in full work, bringing up coal. The witch-hare jumped on to the woodwork which supports the chains, when immediately they refused to move, the engine stopped, pumps, everything remained motionless, and amid the general surprise the witch escaped. But the pit could never be worked again, the pumps and the engine were taken away, and the ruins of the engine house and parts of the other machinery are now pointed out as an undoubted and visible proof of the witch's power.

The witch, being aware of her power over the minds of the people, makes use of it for her own advantage, borrowing her neighbours' horses and farming implements, which they dare not refuse her.

But the most characteristic and general superstition of this part of the country is the "corpse candle." This is seen in various shapes and heard in various sounds; the normal form, from which it takes its name, being, however, a lighted candle, which is supposed to foretell death, by going from the house in which the person dies along the road where the coffin will be carried to the place of burial. It is only a few of the most hardy and best educated who dare to call in question the reality of this fearful omen, and the evidence in support of it is of such a startling and voluminous character, that did we not remember the trials and burnings and tortures for witchcraft and demonianism, and all the other forms of superstition in England but a few years ago, it would almost overpower our common sense.

I will mention a few cases which have been told me by the persons who were witnesses of them, leaving out the hundreds of more marvellous ones which are everywhere to be heard secondhand.

A respectable woman, in a house where we lodged, assured us that on the evening before one of her children died, she saw a lighted candle moving along about three feet from the ground from the foot of the stairs, across the room towards her, that it came close up to her apron and then vanished, and that it was as distinct and plainly visible as the other candles which were in the room.

Another case is of a collier who, going one morning into the pit before any of the other men were at work, heard the coal waggons coming along, although he knew there could be no one then at work. He stood still at the side of the passage, the waggons came along drawn by horses as usual, a man he knew walking in front and another at the side, and the dead body of one of his fellow workmen was in one of the waggons. In the course of the day he related what he had seen to some of the workmen (one of whom told me the story), declaring his belief that the man whose body he had seen would meet with an accident before long. After a year afterwards the man *was* killed by an accident in the pit. The two men seen were near him, and brought him out in the waggon, and their being obliged to stop at the particular place and every other circumstance happened exactly as had been described. This is as the story was told me by a man who declares he heard the prophecy and saw the fulfilment a year afterwards. When such stories are told and believed, it is, of course, useless arguing against the absurdity of it. They naturally say they must believe their own senses, and they are not sufficiently educated to appreciate any general argument you may put to them. There seems to be no fixed time within which the death should follow the "candle" (as all these appearances are called), and therefore when a person sees or thinks he sees anything at night, he sets it down as corpse candle, and by the time he gets home the fright has enlarged it into something marvellously supernatural, and the first corpse that happens to be carried that way is considered to be the fulfilment of it.

There is a general belief that if the person who meets a candle immediately lies down on his back, he will see the funeral procession with every person that will be present, and the corpse with the candle in his hand. There are many strongly authenticated instances of this. One man, on lying down in this manner, saw that it was himself who carried the candle in his hand. He went home, went to bed, never rose from it, but died in a week. These and numberless other stories of a similar character foster the belief in these uneducated people; indeed, it is so general that you can hardly meet a person but can tell you of several marvellous things he has seen himself, besides hundreds vouched for by his neighbours.

They have an account of the origin of this warning in the story of an ancient Welsh bishop, who, while being burnt to death by the Catholics, declared that if his religion was true, a candle should precede every death in the Diocese of St. Davids, going along the exact road the coffin would be carried. They are very incredulous when you tell them that these corpse candles are in great repute in Radnorshire, which is not in the Diocese of St. Davids, and that there are the same appearances under a different name in Ireland.

A celebrated astrologer or conjurer, as he is called in Carmarthenshire, is a living proof of the superstition of the Welsh. This man has printed cards, openly professing to cast nativities, etc., of one of which the following is a literal copy:

"Nativities Calculated,

"In which are given the general transactions of the native through life, viz. Description (without seeing the person), temper, disposition, fortunate or unfortunate in their general pursuits, Honour, Riches, Journeys and Voyages, success therein, and what places best to travel to or reside in; Friends and Enemies, Trade or Profession best to follow and whether fortunate in speculations, viz. Lottery, dealing in foreign markets, &c., &c., &c.

"Of Marriage, if to marry:—The description, temper and disposition of the person; from whence, rich or poor, happy or unhappy in marriage, &c., &c., &c. Of children, whether fortunate or not, &c., &c., &c.

"Deducted from the influence of the Sun and Moon with the Planetary Orbs at the time of birth.

"Also judgment and general issue in sickness, disease, &c. By HENRY HARRIES.

"All letters addressed to him or his father, Mr. JOHN HARRIES, Surgeon, Cwrtycadno, must be post paid or will not be received."

He is, however, most generally consulted when money, horses, sheep, etc., are stolen. He then, without inquiring the time of birth or any other particulars, and without consulting the stars, pretends to know who they are and what they come for. He is, however, generally not at home, and his wife then treats them well, and holds them in conversation till he returns, when he immediately gives them some particulars of the neighbourhood they live in, and pretends to describe the person who stole the goods and the house he lives in, etc., and endeavours to frighten the thief by giving out that he will mark him so that everybody shall know him. In some few cases this succeeds, the person, fearful of the great conjurer's power, returns the goods, and the conjurer then gets great credit. In other cases he manages to tell them something which they cannot tell how he became aware of, and then even if nothing more is heard of the goods, he still keeps up his fame. Two cases have come under my own observation, in which the parties have gone, in one case forty, the other sixty miles, to consult this man about some stolen money; and though in neither case was the desired end obtained, they were told so much about themselves that they felt sure he must have obtained his knowledge by supernatural means. They accordingly spread his name abroad as a wonderful man, who knew a great deal more than other people. The name of his house, "Cwrt y cadno," is very appropriate, as it means in English "The Fox's Court."

Besides these and numberless other instances of almost universal belief in supernatural agency, their superstition as well as their ignorance is further shown by their ascribing to our most harmless reptiles powers of inflicting deadly injury. The toad, newt, lizard, and snake are, they imagine, virulently poisonous, and they look on with horror, and will hardly trust their eyes, should they see them handled with impunity. The barking of dogs at night, hooting of owls, or any unusual noise, dreams, etc., etc., are here, as in many parts of England, regarded as dark omens of our future destiny, mysterious warnings sent to draw aside the veil of futurity and reveal to us, though obscurely, impending danger, disease or death.

≥~ Reckoned by the usual standards on these subjects, the religion of the lower orders of Welshmen may be said to be high in the scale, while their morality is decidedly low. This may appear as a contradiction to some persons, but those who are at all acquainted with mankind well know that, however luxuriantly religion in its outward forms and influence on the tongue may flourish in an uncultivated soil, it is by no means necessarily accompanied by an equal growth of morality. The former, like the flower of the field, springs spontaneously, or with but little care; the latter, like the useful grain, only by laborious cultivation and the careful eradication of useless or noxious weeds.

If the number of chapels and prayer-meetings, the constant attendance on them, and the fervour of the congregation can be accounted as signs of religion, it is here. Besides the regular services on the Sabbath and on other days, prayer-meetings are held early in the morning and late at night in different cottages by turns, where the uneducated agriculturist or collier breathes forth an extempory prayer. The Established Church is very rarely well attended. There is not enough of an exciting character or of originality in the service to allure them, and the preacher is too frequently an Englishman who speaks the native tongue, but as a foreigner.

Their preachers, while they should teach their congregation moral duties, boldly decry their vices, and inculcate the commandments and the duty of doing to others as we would they should do unto us, here, as is too frequently the case throughout the kingdom, dwell almost entirely on the mystical doctrine of the atonement—a doctrine certainly not intelligible to persons in a state of complete ignorance, and which, by teaching them that they are not to rely on their own good deeds, has the effect of entirely breaking away the connection between their religion and the duties of their everyday life, and of causing them to imagine that the animal excitement which makes them groan and shriek and leap like madmen in the place of worship, is the true religion which will conduce to their happiness here, and lead them to heavenly joys in a world to come.

Among the youth of both sexes, however, the chapel and prayer-meeting is considered more in the light of a "trysting" place than as a place of worship, and this is one reason of the full attendance especially at the evening services. And as the meetings are necessarily in a thinly populated country, often distant, the journey, generally performed on horseback, affords opportunities for converse not to be neglected.

Thus it will not be wondered at, even by those who affirm the connection between religion and morality, that the latter is, as I said before, at a very low ebb. Cheating of all kinds, when it can be done without being found out, and all the lesser crimes are plentiful enough. The notoriety which Welsh juries and Welsh witnesses have obtained (not unjustly) shows how little they scruple to break their word or oath. Having to give their evidence through the medium of an interpreter gives them the advantage in court, as the counsel's voice and manner have not so much effect upon them. They are, many of them, very good witnesses as far as sticking firmly to the story they have been instructed in goes, and returning the witticisms of the learned counsel so as often to afford much mirth. To an honest jury a Welsh case is often very puzzling, on account of its being hardly possible to get a single fact but what is sworn against by an equal number of witnesses of the opposite side; but to a Welsh jury, who have generally decided on their verdict before the trial commences, it does not present any serious difficulty.

The morals and manners of the females, as might be expected from entire ignorance, are very loose, and perhaps in the majority of cases a child is born before the marriage takes place.

But let us not hide the poor Welshman's virtues while we expose his faults. Many of the latter arise from his desire to defend his fellow countrymen from what he considers unfair or unjust persecution, and many others from what he cannot himself prevent—his ignorance. He is hospitable even to the Saxon, his fire, jug of milk, and bread and cheese being always at your service. He works hard and lives poorly. He bears misfortune and injury long before he complains. The late Rebecca disturbances, however, show that he may be roused, and his ignorance of other effectual measures should be his excuse for the illegal and forcible means he took to obtain redress—means which, moreover, have been justified by success. It is to be hoped that he will not have again to resort to such outrages as the only way to compel his rulers to do him justice.

A broader system of education is much needed in the Principality. Almost all the schools, it is true, teach the English language, but the child finds the difficulty of acquiring even the first rudiments of education much increased by his being taught them in an unfamiliar tongue of which he has perhaps only picked up a few commonplace expressions. In arithmetic, the new language presents a greater difficulty, the method of enumerating being different from their own; in fact, many Welsh children who have been to school cannot answer a simple question in arithmetic till they have first translated it into Welsh. Unless, therefore, they happen to be thrown among English people or are more than usually well instructed, they get on but little with anything more than speaking English, which those who have been to school generally do very well. Whatever else they have learnt is soon lost for want of practice. It would be very useful to translate some of the more useful elementary works in the different branches of knowledge into Welsh, and sell them as cheaply as possible. The few little Welsh books to be had (and they are very few) are eagerly purchased and read with great pleasure, showing that if the means of acquiring knowledge are offered him, the Welshman will not refuse them.

I will now conclude this brief account of the inhabitants of so interesting a part of our island, a part of which will well repay the trouble of a visit, as much for its lovely vales, noble mountains, and foaming cascades, as for the old customs and still older language of the inhabitants of the little white-washed cottages which enliven its sunny vales and barren mountain slopes.

At Neath

At Easter I bade farewell to Leicester and went to Neath with my brother John, in order to wind up our brother William's affairs. We found from his books that a considerable amount was owing to him for work done during the past year or two, and we duly made out accounts of all these and sent them in to the respective parties. Some were paid at once, others we had to write again for and had some trouble to get paid. Others, again, were disputed as being an extravagant charge for the work done, and we had to put them in a lawyer's hands to get settled. One gentleman, whose account was a few pounds, declared he had paid it, and asked us to call on him. We did so, and, instead of producing the receipt as we expected, he was jocose about it, asked us what kind of business men we were to want him to pay twice; and when we explained that it was not shown so in my brother's books, and asked to look at the receipt, he coolly replied, "Oh, I never keep receipts; never kept a receipt in my life, and never was asked to pay a bill twice till now!" In vain we urged that we were bound as trustees for the rest of the family to collect all debts shown by my brother's books to be due to him, and if he did not pay it, we should have to lose the amount ourselves. He still maintained that he had paid it, that he remembered it distinctly, and that he was not going to pay it twice. At last we were obliged to tell him that if he did not pay it we *must* put it in the hands of a lawyer to take what steps he thought necessary; then he gave way, and said, "Oh, if you are going to law about such a trifle, I suppose I must pay it again!" and, counting out the money, added, "There it is; but I paid it before, so give me a receipt *this* time," apparently considering himself a very injured man. This little experience annoyed me much, and, with others of the same nature later on, so disgusted me with business as to form one of the reasons which induced me to go abroad.

When we had wound up William's affairs as well as we could, my brother

My Life, 1:241–49.

John returned to London, and I was left to see if any work was to be had, and in the mean time devoted myself to collecting butterflies and beetles. While at Leicester I had been altogether out of the business world, and do not remember even looking at a newspaper, or I might have heard something of the great railway mania which that year reached its culmination. I now first heard rumours of it, and someone told me of a civil engineer in Swansea who wanted all the surveyors he could get, and that they all had two guineas a day, and often more. This I could hardly credit, but I wrote to the gentleman, who soon after called on me, and asked me if I could do levelling. I told him I could, and had a very good level and levelling staves. After some little conversation he told me he wanted a line of levels up the Vale of Neath to Merthyr Tydfil for a proposed railway, with cross levels at frequent intervals, and that he would give me two guineas a day, and all expenses of chain and staff men, hotels, etc. He gave me all necessary instructions, and said he would send a surveyor to map the route at the same time. This was, I think, about mid-summer, and I was hard at work till the autumn, and enjoyed myself immensely. It took me up the south-east side of the valley, of which I knew very little, along pleasant lanes and paths through woods and by streams, and up one of the wildest and most picturesque little glens I have ever explored. Here we had to climb over huge rocks as big as houses, ascend cascades, and take cross-levels up steep banks and precipices all densely wooded. It was surveying under difficulties, and excessively interesting. After the first rough levels were taken and the survey made, the engineers were able to mark out the line provisionally, and I then went over the actual line to enable the sections to be drawn as required by the Parliamentary Standing Orders.

In the autumn I had to go to London to help finish the plans and reference books for Parliament. There were about a dozen surveyors, draughtsmen, and clerks in a big hotel in the Haymarket, where we had a large room upstairs for work, and each of us ordered what we pleased for our meals in the coffee-room. Towards the end of November we had to work very late, often till past midnight, and for the last few days of the month we literally worked all night to get everything completed.

In this year of wild speculation it is said that plans and sections for 1263 new railways were duly deposited, having a proposed capital of £563,000,000, and the sum required to be deposited at the Board of Trade was so much larger than the total amount of gold in the Bank of England and notes in circulation at the time, that the public got frightened, a panic ensued, shares in the new lines which had been at a high premium fell almost to nothing, and even the established lines were greatly depreciated. Many of the lines were proposed merely for speculation, or to be bought off by opposing lines which had a better chance of success. The line we were at work on was a branch of the Great Western and South Wales Railway then making, and was for the purpose of bringing the coal and iron of Merthyr Tydfil and the surrounding district to Swansea, then the chief port of South Wales. But we had a competitor along the whole of our route in a great line from Swansea to Yarmouth, by way of Merthyr, Hereford, Worcester, and across the midland agricultural counties, called, I think, the East and West Junction Railway, which sounded grand, but which had no chance of passing. It competed, however, with several other lines, and I heard that many of these agreed to make up a sum to buy off its opposition. Not one-tenth of the lines proposed that year were ever made, and the money wasted upon surveyors, engineers, and law expenses must have amounted to millions.

Finding it rather dull at Neath living by myself, I persuaded my brother to give up his work in London as a journeyman carpenter and join me, thinking that, with his practical experience and my general knowledge, we might be able to do architectural, building, and engineering work, as well as surveying, and in time get up a profitable business. We returned together early in January, and continued to board and lodge with Mr. Sims in the main street, where I had been very comfortable, till the autumn, when, hearing that my sister would probably be home from America the following summer, and my mother wishing to live with us, we took a small cottage close to Llantwit Church, and less than a mile from the middle of the town. It had a nice little garden and yard, with fowl-house, shed, etc., going down to the Neath Canal, immediately beyond which was the river Neath, with a pretty view across the valley to Cadoxton and the fine Drumau Mountain.

Having the canal close at hand and the river beyond, and then another canal to Swansea, made us long for a small boat, and not having much to do, my brother determined to build one, so light that it could easily be drawn or carried from the canal to the river, and so give access to Swansea. It was made as small and light as possible to carry two or, at most, three persons. When finished, we tried it with much anxiety and found it rather unstable, but with a little ballast at the bottom and care in moving, it did very well, and was very easy to row. One day I persuaded my mother to let me row her to Swansea, where we made a few purchases; and then came back quite safely till within about a mile of home when, passing under a bridge, my mother put her hand out to keep the boat from touching, and leaning over a little too much, the side went under water, and upset us both. As the water was only about two or three feet deep we escaped with a thorough wetting. The boat was soon bailed dry, and then I rowed on to Neath Bridge, where my mother got out and walked home, and did not trust herself in our boat again, though I and my brother had many pleasant excursions.

Our chief work in 1846 was the survey of the parish of Llantwit-juxta-Neath, in which we lived. The agent of the Gnoll Estate had undertaken the valuation for the tithe commutation, and arranged with me to do the survey and make the map and the necessary copies. When all was finished and the valuation made, I was told that I must collect the payment from the various farmers in the parish, who would afterwards deduct it from their rent. This was a disagreeable business, as many of the farmers were very poor; some could not speak English, and could not be made to understand what it was all about; others positively refused to pay; and the separate amounts were often so small that it was not worth going to law about them, so that several were never paid at all, and others not for a year afterwards. This was another of the things that disgusted me with business, and made me more than ever disposed to give it all up if I could but get anything else to do.

We also had a little building and architectural work. A lady wanted us to design a cottage for her, with six or seven rooms, I think, for £200. Building with the native stone was cheap in the country, but still, what she wanted was impossible, and at last she agreed to go £250, and with some difficulty we managed to get one built for her for this amount. We also sent in a design for a new Town Hall for Swansea, which was beyond our powers, both of design and draughtsmanship; and as there were several established architects among the competitors, our very plain building and poor drawings had no chance. But shortly afterwards a building was required at Neath for a Mechanics' Institute, for which £600 was available. It was to be in a narrow side street, and to consist of two rooms only, a reading room and library below, and a room above for classes and lectures. We were asked to draw the plans and supervise the execution, which we did, and I think the total cost did not exceed the sum named by more than £50. It was, of course, very plain, but the whole was of local stone, with door and window-quoins, cornice, etc., hammer-dressed; and the pediments over the door and windows, arched doorway, and base of squared blocks gave the whole a decidedly architectural appearance. It is now used as a free library, and through the kindness of Miss Florence Neale, of Penarth, I am enabled to give a photographic reproduction of it [Figure 4].

This reminds me that the Mechanics' Institution was, I think, established by Mr. William Jevons, a retired merchant or manufacturer of Liverpool, and the uncle of William Stanley Jevons, the well-known writer on Logic and Political Economy. Mr. Jevons was the author of a work on "Systematic Morality," very systematic and very correct, but as dry as its title. He had a good library, and was supposed in Neath to be a man of almost universal knowledge. I think my brother William had become acquainted with him after I left Neath, as he attended the funeral, and I and John spent the evening with him. When I came to live in Neath after my brother's death, I often saw him and occasionally visited him, and I think borrowed books, and the following winter, finding I was interested in science generally, he asked me to give some familiar lectures or lessons to the mechanics of Neath, who then met, I think, in one of the schoolrooms. I was quite afraid of undertaking this, and tried all I could to escape, but Mr. Jevons was very persistent, assured me that they knew actually nothing of science, and that the very simplest things, with a few diagrams and experiments, would be sure to interest them. At last I reluctantly consented, and begun with very short and simple talks on the facts and laws of mechanics, the principle of the lever, pulley, screw, etc., falling bodies and projectiles, the pendulum, etc.

I got on fairly well at first, but on the second or third occasion I was trying to explain something which required a rather complex argument which I thought I knew perfectly, when, in the middle of it, I seemed to lose myself and could not think of the next step. After a minute's dead silence, Mr. Jevons, who sat by me, said gently—"Never mind that now. Go on to the next subject." I did so, but after a few minutes, what I had forgotten became clear to me, and I returned to it, and went over it with success. I gave these lessons for two winters, going through the elementary portions of physics; and after a week in Paris in 1847, I gave to the same audience a general account of the city, with special reference to its architecture, museums, and gardens, showing that it was often true that "they did these things better in France."*

There was also in Neath a Philosophical Society with a small library and reading room, in connection with which occasional lectures were given. Sir G. B. Airy, the Astronomer Royal, gave a lecture there on the return of Halley's Comet shortly before we came to Neath. He recommended them to purchase a good telescope of moderate size and have it properly mounted, so as to be able to observe all the more remarkable astronomical phenomena. A telescope was actually obtained with, I think, a four- or five-inch object glass, and as there was no good position for it available, a kind of square tower was built attached to the library, high enough to obtain a clear view, on the top of which it was proposed to use the telescope. But the funds for a proper mounting and observatory roof not being forthcoming, the telescope was hardly ever used, owing to the time and trouble always required to carry upstairs and prepare for observation any astronomical telescope above the very smallest size.

✍ During the two summers that I and my brother John lived at Neath we spent a good deal of our leisure time in wandering about this beautiful district, on my part in search of insects, while my brother always had his eyes open for any uncommon bird or reptile. One day when I was insect hunting on Crymlyn Burrows, a stretch of very interesting sand-hills, rock, and bog near the sea, and very rich in curious plants, he came upon several young vipers basking on a rock. They were about eight or nine inches long. As they were quite still, he thought he could catch one by the neck, and endeavoured to do so, but the little creature turned round suddenly, bit his finger, and escaped. He immediately sucked out the poison, but his whole hand swelled considerably, and was very painful. Owing, however, to the small size of the animal the swelling soon passed off, and left no bad

---

* In 1895 I received a letter from Cardiff, from one of the workmen who attended the Neath Mechanics' Institution, asking if the author of "Island Life," the "Malay Archipelago," and other books is the same Mr. Alfred Wallace who taught in the evening science classes to the Neath Abbey artificers. He writes—"I have often had a desire to know, as I benefited more while in your class—if you are the same Mr. A. Wallace—than I ever was taught at school. I have often wished I knew how to thank you for the good I and others received from your teaching.—(Signed) MATTHEW JONES."

effects. Another day, towards the autumn, we found the rather uncommon black viper in a wood a few miles from Neath. This he caught with a forked stick, to which he then tied it firmly by the neck, and put it in his coat pocket. Meeting a labourer on the way, he pulled it out of his pocket, wriggling and twisting around the stick and his hand, and asked the man if he knew what it was, holding it towards him. The man's alarm was ludicrous. Of course, he declared it to be deadly, and for once was right, and he added that he would not carry such a thing in his pocket for anything we could give him.

Though I have by no means a very wide acquaintance with the mountain districts of Britain, yet I know Wales pretty well; have visited the best parts of the lake district; in Scotland have been to Loch Lomond, Loch Katrine, and Loch Tay; have climbed Ben Lawers, and roamed through Glen Clova in search of rare plants; but I cannot call to mind a single valley that in the same extent of country comprises so much beautiful and picturesque scenery, and so many interesting special features, as the Vale of Neath. The town itself is beautifully situated, with the fine wooded and rockgirt Drumau Mountain to the west, while immediately to the east are well-wooded heights crowned by Gnoll House, and to the south-east, three miles away, a high rounded hill, up which a chimney has been carried from the Cwm Avon copperworks in the valley beyond, the smoke from which gives the hill much the appearance of an active volcano. To the south-west the view extends down the valley to Swansea Bay, while to the north-east stretches the Vale of Neath itself, nearly straight for twelve miles, the river winding in a level fertile valley about a quarter to half a mile wide, bounded on each side by abrupt hills, whose lower slopes are finely wooded, and backed by mountains from fifteen hundred to eighteen hundred feet high. The view up this valley is delightful, its sides being varied with a few houses peeping out from the woods, abundance of lateral valleys and ravines, with here and there the glint of falling water, while its generally straight direction affords fine perspective effects, sometimes fading in the distance into a warm yellow haze, at others affording a view of the distant mountain ranges beyond. . . .

～ 2 ～

The Amazon

Wallace and Bates chose to go to Brazil to collect animals after reading W. H. Edwards's book, *A Voyage Up the Amazon* (1847).[1] In addition to describing the beauty of tropical vegetation, Edwards gave an encouraging account of the hospitality of the local people and the moderate expenses of living and traveling in Brazil. The amateur naturalists met Mr. Edwards by chance while they were in London making arrangements for their trip, and he wrote letters of introduction for them to his American friends in Pará. They sought advice from British Museum entomologist Mr. Edward Doubleday, who assured them that because of the extensive London trade in exotic insects, shells, and birds, they could earn enough money to pay their way. Wallace and Bates also obtained letters of introduction from a butterfly collector in London who had been to Brazil and from the captain of the ship on which they sailed. They arrived in Pará on May 28, 1848, after a month-long sea journey. They acquainted themselves with their new surroundings and began their first collecting excursions from their temporary base in Pará at the country house of Mr. Miller, brother of the ship's captain.

Wallace had been in Brazil for more than a year when his younger brother Herbert arrived in July 1849 to assist in collecting natural history specimens (Herbert died in 1851). The same ship that brought Herbert also brought British botanist Richard Spruce to Pará. Shortly after, the Wallace brothers and Henry Bates spent several months collecting in and around Santarem with Richard Spruce and his assistant Robert King. Wallace, Bates, and Spruce, contemporaries in Amazonia in the mid-nineteenth century, formed a somewhat unusual triumvirate of self-taught naturalists who made their way into mainstream British natural science. This was one of several times that Wallace and Spruce spent time together in Brazil; they had many interests in common and formed a friendship that would last for many decades. Wallace later wrote that Spruce was "among the dearest of my friends, the one towards whom I felt more like a brother than to any other person."[2] And it was in conversations with, and letters to, Bates that young Wallace candidly spelled out his ideas about the origins of plants and animals. They found kindred spirits in one another as each struggled to make a career in natural history and to survive the hardships of tropical collecting.

These friendships forged and tested in the Amazonian tropics were enhanced by common interests in nature no less than by the arduous, at times life-threatening, circumstances. As Wallace explored the regions along the Rio Negro, he reached the village of San Carlos, Venezuela, the furthest point reached by Humboldt and Bonpland from the opposite direction some fifty years earlier (Figure 5). At San Carlos Wallace collected three of the several new species of palm he described. As he continued his explorations of northern Amazonia, he found traveling up the falls of the Rio Uaupés extremely difficult but particularly promising for his bird collection. He decided on the spot to change his plan for a trip to Andes: he would return with a better canoe and more men for a second journey further up the Uaupés, abandon his idea of going to the Andes, and return to England after this planned river journey. Having made his decision, he realized how much he longed for home, remarking in his narrative that his yearning for the tropics had never felt so powerful as his anticipation of returning to familiar faces, foods, and flowers. Yet the intensity of his curiosity bade him stay for one more river journey, lured by the possibility of finding evidence for his hypothesis of common descent, believing that the most closely related species would be found near one another.[3]

As Wallace prepared for his last Amazonian river adventure, he not only lost one after another of his men to sickness, but he too contracted

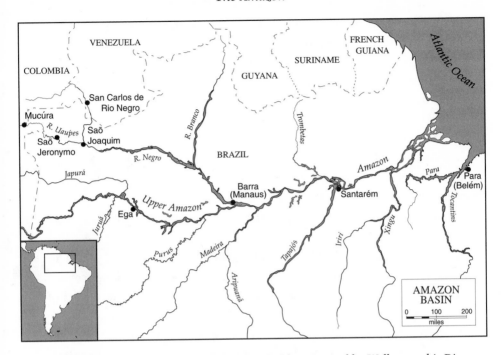

Fig. 5. Map of the Amazon basin showing the places visited by Wallace on his Rio Negro journey. The place names are those used by Wallace; present names for Pará and Barra are in parentheses. Prepared by the University of Wisconsin Cartographic Laboratory.

yellow fever. His trip up the Uaupés had to be delayed, in spite of his desire to return there for the dry season, the most favorable for collecting. A friend who nursed him, Senhor João Antonio de Lima, a Brazilian businessman, did not expect Wallace to survive, so intense were the fits of fever. Having been alerted to Wallace's situation, Spruce came to see him, and after several months, Wallace was well enough to return the visit. It is at this point that the excerpt from his narrative of his second journey up the Uaupés begins.

Note that as the chapter opens, Wallace is still weak but decides that he may just as well be confined to a canoe as to a village house. It is clear from this portion of his book, as well as the rest of the narrative, that the Indians were just as interesting to him, just as much a part of the landscape he witnessed, as were the plants and animals. This journey up the Uaupés was particularly successful, as it provided Wallace with measurements of the course of the river, which, combined with his observations

of the Rio Negro, gave him unique knowledge of the geography of a remote and unmapped region of the Amazon. The selection here provides
the flavor of much of his journey by relating the almost unrelenting details of finding and losing Indians to help man the canoe and collect animals, the difficulties of procuring food for the live monkeys and birds he
hoped to bring back to England with him, and the thrill of being where
no European had been before. Notice too that for Wallace, as for most
nineteenth-century naturalists, the hunting, killing, and collecting of animals and plants for the specimen trade and for the pursuit of science
were a necessary part of their job.

The next selection is Wallace's letter to Spruce detailing the disastrous
trip back to England; the letter describing the nearly tragic journey
speaks for itself. At the end of the same year of his return, 1852, Wallace
read a paper at the December 14 meeting of the Zoological Society of
London. His brief, nontechnical descriptions of the twenty-one species of
monkeys he observed on his journey are followed by a few remarks on
their geographical distribution. What stands out is a distinct lament
about the lack of geographical precision in existing museum collections
and printed works. His short paper is reproduced in its entirety as the final selection in this chapter, but several sentences bear repetition because they show Wallace's willingness to criticize existing collections and
his indirect allusion to ancestral relationships among species, that is,
common descent. "Owing to this uncertainty of locality, and the additional confusion created by mistaking allied species from distant countries, there is scarcely an animal whose exact geographical limits we can
mark out on the map. On this accurate determination of an animal's
range many interesting questions depend. Are very closely allied species
ever separated by a wide interval of country?"

In one of his most important scientific papers just three years later, the
"Law" paper of 1855, Wallace would answer his own question, arguing
convincingly that closely allied species are found in closely adjoining locations and in the same geological period because one has descended
from the other. This was a crucial building block in his understanding of
evolutionary processes. In his article on the monkeys of the Amazon, we
can see the rudiments of what would become the pattern of much of his
later zoological research, an emphasis on mapping and on precise observations of geographical distribution, including both the type of habitat

and the size of an organism's range. He saw geographical distribution as a fundamental characteristic of living forms and as crucial evidence for understanding their evolutionary history. In emphasizing distribution, Wallace was a major player in the shift from descriptive natural history to a more analytic approach in studying organisms.

We might compare the study of plant and animal geography in the nineteenth century with ecology or environmental studies in the twentieth: it was *the* new synthesis, it was natural history coming of age, asking some of the most difficult and compelling questions about the nature and history of life on earth. Beginning with the work of continental scientists such as Eberhardt von Zimmermann, Augustin de Candolle, and Alexander von Humboldt in the late eighteenth and early nineteenth centuries, studies of the geographical distribution of plants and animals became increasingly prominent in European and American scientific publications. Darwin devoted two entire chapters of the *Origin of Species* (1859) to geographical distribution, the only subject to which he gave such lengthy attention. Distribution maps rarely appeared in school geographies and popular atlases before 1850, but after mid-century they became more and more commonplace. During the last third of the century, people working in quite varied kinds of natural history depended on distribution mapping. Wallace later authored one of the great syntheses in this field, *The Geographical Distribution of Animals* (1876). In the paper on monkeys, the amateur Wallace is straightforward, even bold, in his admonition to other zoologists to pay closer attention to exactly how animals are distributed geographically.

Although the written results of his trip were severely compromised by the loss of his collections, Wallace's emphasis on geographical distribution is evident in practically all of the publications from this period. Some of his observations about plant and animal distributions are included on the map he drew of his explorations on the Rio Negro and the Rio Uaupés. The hand-drawn and colored map he made when he arrived back in England impressed the fellows and officers of the Royal Geographical Society, and for many years it was the only map of this region of the Amazon basin. The original map is large, 84 × 130 cm (approximately 33 × 51 inches), and shows great care and skill in its execution. A lithographed and much reduced version of Wallace's hand-drawn map accompanied the published version of his descriptions of the rivers (Figure

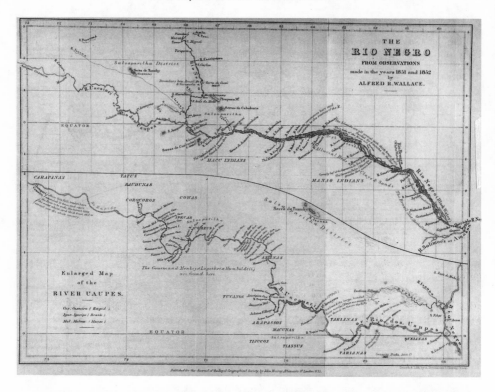

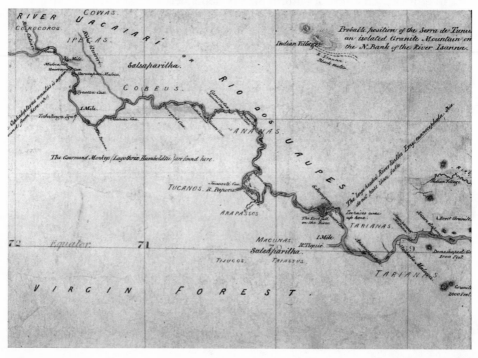

6). Although the reduced, published version contains all of the information from Wallace's original, it was divided into two parts: the upper portion shows the extent of the Rio Negro, including the Rio Uaupés; the lower portion is an enlarged map of the Uaupés, with additional place names and descriptive notes. Because it is reduced even further here, it is quite difficult to read the place names and notes about the distribution of plants and animals. A small portion of Wallace's manuscript map, at close to the original size, is shown in Figure 7, which has the graceful style of the original map. A general map of the region is included to help locate the places Wallace explored and mapped (Figure 5).

~❧ ❦~

On the Rio Negro

At length, on the 16th of February, two months and twenty-three days after my arrival at São Joaquim, I left on my voyage up the Uaupés. I was still so weak that I had great difficulty in getting in and out of the canoe; but I thought I should be as well there as confined in the house; and as I now

A Narrative of Travels on the Amazon and Rio Negro, chap. 12.

FACING PAGE:

Top: Fig. 6. Wallace's lithographed map of Rio Negro. It was not prepared in time to be included in Wallace's narrative of his Amazon journey, but it shows the main route he traveled and highlights the newly explored region of the Uaupés in the enlarged map on the bottom. From "On the Rio Negro," *Journal of the Royal Geographical Society of London,* 23 (1853):212–17; facing p. 212.

Bottom: Fig. 7. Portion of Wallace's original map of the Rio Negro showing the Rio Uaupés. Wallace's concern with distribution is shown in the several descriptive notes about animals. Some notes indicate range limits, such as "the large headed River turtles do not pass these falls," whereas others indicate areas of distribution, such as "the Gourmand Monkeys are found here." The note on the upper left, "the *Cephalopterus ornatus* is again found from here up," refers to the umbrella bird. The abbreviation "Cax." stands for *caxoeira,* or rapids, of which the more difficult ones are named. Wallace's original map is mounted on linen and housed at the Royal Geographical Society of London.

longed more than ever to return home, I wished first to make this voyage, and get a few living birds and animals to take with me. I had seven Uaupés Indians that Senhor L. [Senhor João Antonio de Lima, a Brazilian businessman based in São Joaquim, who helped Wallace on his Rio Negro journeys] had brought from São Jeronymo, in order to take me up the river. Three more, who had already received payment for the voyage, did not appear; and, though they knew very well the time of my leaving, had fixed on that very day to give a feast of fish and caxirí. Antonio, my former pilot to Barra, was one. I met him coming to the village from his sitio [small farm], and he flatly refused to come with me, unless I waited some days more for him; I therefore made him send his Macu boy, João, instead, to go and return, and so pay for what both owed. This he did, and we went on our way rejoicing, for Antonio was what they call an Indian "ladino," or crafty; he could speak Portuguese, and, strongly suspecting him of being an expert thief, I was not sorry to be without his company.

On Saturday evening, the 21st, we arrived at São Jeronymo, where I was cordially received by Senhor Augustinho. The next day was occupied in paying my men, and sending for Bernardo to conduct my canoe up the falls, and get me more Indians for the voyage.

On Monday he arrived, and I let him take the canoe, but did not go with him, as, for some days past, the ague had again attacked me, and this was the day of the fit; so I sent the two guardas, my head men, who could speak Portuguese, to take charge of the canoe and cargo, and remained myself till the next day. In the evening a small trader arrived from above, very tipsy, and an Indian informed Senhor Augustinho that it was with my caxaça [strong liquor made from sugarcane], which the men whom I had brought specially to take charge of my cargo, had opened. This I next day found to be the case, as the seals had been broken, and clumsily refastened with a burning stick. These men were half-civilised Indians, who came with me as hunters, to interpret for me with the Indians and take charge of my goods, on account of which I paid them extra wages. They ate with me, and did not row with the other Indians; but the temptation of being left along for nearly a day, with a garafão of caxaça, was too strong for them. Of course I passed all over in silence, appearing to be perfectly ignorant of what had taken place, as, had I done otherwise, they would probably both have left me, after having received the greater part of their payment beforehand, and I should have been unable to proceed on my voyage.

With Bernardo's assistance, I soon got ten paddles in my canoe; and having paid most of them out of my stock of axes, mirrors, knives, beads, etc., we went along very briskly to Jauarité, where we arrived on the morning of the 28th. I was anxious to pass the caxoeira [cascades, rapids, falls] immediately, but was delayed,—paying two Indians, who left me here, and procuring others; so my ague fit fell upon me before we left the village, and I was very weak and feverish when we went to pass the falls. We unloaded the whole of the cargo, which had to be carried a considerable distance through the forest; and even then, pulling the canoe up the falls was a matter of great difficulty. There are two falls, at some distance from each other, which make the land-carriage very long.

We then re-embarked, when Bernardo coolly informed me that he could go no further, after having received payment for the whole voyage. His brother, he said, should go in his place; and when I returned, he would pay me what he owed me. So I was forced to make the best of it; but shortly after I found that his brother would only go to Jacaré caxoeira, and thus I was a second time deceived.

On starting, I missed João, and found that he had left us in the village, telling the guardas that he had only agreed with me to come so far, and they had never said a word to me about it till now, that it was too late. Antonio's debt therefore still remained unpaid, and was even increased by a knife which João had asked for, and I had given him, in order that he might go on the voyage satisfied.

The river now became full of rocks, to a degree to which even the rockiest part of the Rio Negro was a trifle. All were low, and would be covered at high-water, while numbers more remained below the surface, and we were continually striking against them. That afternoon we passed four more falls, the "Uacú" (a fruit), "Uacará" (Egret), "Mucúra" (Opossum), and "Japóna" (oven) caxoeiras. At Uacará there was a malocca of the same name; and at Japóna another, where we passed the night [Figure 8]. All these rapids we ascended without unloading; but the Uacará was very bad, and occasioned us much trouble and delay. The next morning, when about to start, we found that another Indian was missing: he had absconded in the night, and it was useless attempting to seek him, though we knew he had gone to Uacará Malocca, where he wished to stay the day before, but where all knowledge of him would be denied and he well hidden, had we returned to fetch him. He was one who had received full payment, making

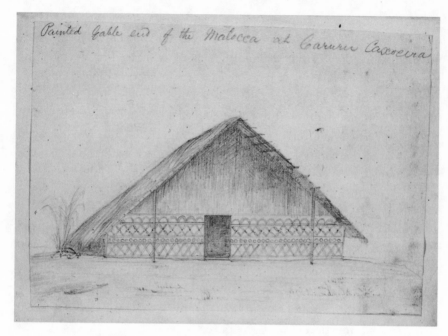

Fig. 8. Wallace's pencil drawing of a *malocca*, a native village house, on the Rio Uaupés. From the Linnean Society of London.

three who had already gone away in my debt; a not very encouraging beginning for my voyage.

We passed the "Tyeassu" (Pig) caxoeira early, and then had a good stretch of quiet water till midday, when we reached the "Oomarie" (a fruit) caxoeira, where there is a sitio. Here we dined off a fine fresh Tucunaré, which an old man sold me; and I agreed with his son, by the temptation of an axe, to go with me. We pulled the canoe up this rapid without unloading, which is seldom done, except when the river is low, as it now was. The rest of the day we had quiet water, and stopped at a rock to make our supper and sleep.

March 1st.—We passed the "Macáco" (monkey) caxoeira early. The rocks here, and particularly about Oomarie caxoeira, were so full of parallel veins, as to give them the appearance of being stratified and thrown up nearly vertically; whereas they are granitic, and similar to those we had already seen. We then soon reached the "Irá" (Honey) and "Baccába" (a Palm) caxoeiras; at both of which there are figures or picture-writings on the rocks, which I stayed to sketch. In passing the latter rapid, we knocked off one of the false keels I had had put to the canoe previous to starting, to preserve the bottom in the centre, where it was worn very thin by being dragged over the rocks by its former owner. We therefore stopped at a sandbank, unloaded the canoe, and plugged up the nail-holes, which were letting in water very fast.

The next day we passed in succession the "Arára Mirí" (Little Macaw), "Tamaquerié" (Gecko), "Paroquet," "Japoó" (a bird), "Arára" (Macaw), "Tatú" (Armadillo), "Amána" (Rain), "Camóa" (?), "Yauti" (Tortoise); and, finally, about three P.M., arrived at "Carurú" (a water-plant) caxoeira. The last five of these, before arriving at Carurú, were exceedingly bad; the passage being generally in the middle of the river, among rocks, where the water rushes furiously. The falls were not more than three or four feet each; but, to pull a loaded canoe up these, against the foaming waters of a large river, was a matter of the greatest difficulty for my dozen Indians, their only resting-place being often breast-deep in water, where it was a matter of wonder that they could stand against the current, much less exert any force to pull the canoe. At Arára fall, the usual passage is over the dry rock, and we unloaded for that purpose; but all the efforts of the Indians could not get the heavy canoe up the steep and rugged ascent which was the only pathway. Again and again they exerted themselves, but to no purpose; and I was just sending by an old man, who was passing in a small canoe, to Carurú for assistance, when he suggested that by getting a long sipó [rope made from the stem of a plant] (the general cable in these rivers) we might obtain a good purchase, to pull the canoe up the margin of the fall, which we had previously tried without success. We accordingly did so, and by great exertions the difficulty was passed,—much to my satisfaction, as sending to Carurú would have occasioned a great and very annoying delay.

The river from Jauarité may be said to average about a third of a mile wide, but the bends and turns are innumerable; and at every rapid it almost always spreads out into such deep bays, and is divided into channels by so many rocks and islands, as to make one sometimes think that the water is suddenly flowing back in a direction contrary to that it had previously been taking. Carurú caxoeira itself is greater than any we had yet seen,—rushing amongst huge rocks down a descent of perhaps fifteen or twenty feet. The only way of passing this, was to pull the canoe over the dry rock, which rose considerably above the level of the water, and was rather rugged, being interrupted in places by breaks or steps two or three feet high. The canoe was accordingly unloaded, quantities of poles and branches cut and laid in the path to prevent the bottom being much injured by the rocks, and a messenger sent to the village on the other side of the river to request the Tushaúa to come with plenty of men to our assistance. He soon arrived with eleven Indians, and all hands set to work pushing the canoe, or pulling at the sipós; and even then, the strength of five-and-twenty persons could only move it by steps, and with great difficulty. However, it was at length passed, and we then proceeded to the village, where the Tushaúa lent us a house.

The canoe was so weak in the bottom in one place, that I was fearful of some accident in my descent, so I determined to stay here two or three days, to cut out the weak part and put in a strong board. I now also saw that this canoe was much too heavy to proceed further up the river, as at many of the falls there was no assistance to be obtained, even in places as difficult to pass as Carurú; so I opened negotiations to purchase a very large "obá" [a dugout canoe (a pirogue) made by burning and digging out a tree trunk] of the Tushaúa, which, before leaving, I effected for an axe, a shirt and trousers, two cutlasses, and some beads. We were delayed here five entire days, owing to the difficulty of finding a tree of good wood sufficiently large to give a board of twelve or fourteen inches wide; and at last I was obliged to be content with two narrow boards, clumsily inserted, rather than be exposed to more delay.

There was a large malocca here, and a considerable number of houses [see Figure 8]. The front of the malocca was painted very tastefully in diamonds and circles, with red, yellow, white, and black. On the rocks were a series of strange figures, of which I took a sketch. The Indians were of the "Ananás" or Pineapple tribe; I bought some dresses and feather ornaments of them; and fish, mandiocca-cakes, etc., were brought me in considerable quantities, the articles most coveted in return being fish-hooks and red beads, of both of which I had a large stock. Just below the fall, the river is not more than two or three hundred yards wide; while above, it is half a mile, and contains several large islands.

The large black pacu was abundant here, and, with other small fish, was generally brought us in sufficient quantity to prevent our recurring to fowls, which are considered by the traders to be the most ordinary fare a man can live on. I now ate for the first time the curious river-weed, called carurú, that grows on the rocks. We tried it as a salad and also boiled with fish; and both ways it was excellent;—boiled, it much resembled spinach.

Here, too, I first saw and heard the "Juriparí," or Devil-music of the Indians. One evening there was a caxirí-drinking; and a little before dusk a sound as of trombones and bassoons was heard coming on the river towards the village, and presently appeared eight Indians, each playing on a great bassoon-looking instrument. They had four pairs, of different sizes, and produced a wild and pleasing sound. They blew them all together, tolerably in concert, to a simple tune, and showed more taste for music than I had yet seen displayed among these people. The instruments are made of bark spirally twisted, and with a mouthpiece of leaves.

In the evening I went to the malocca, and found two old men playing on the largest of the instruments. They waved them about in a singular manner, vertically and sideways, accompanied by corresponding contortions of the body, and played a long while in a regular tune, accompanying each other very correctly. From the moment the music was first heard, not a female, old or young, was to be seen; for it is one of the strangest superstitions of the Uaupés Indians, that they consider it so dangerous for a woman ever to see one of these instruments, that having done so is punished with death, generally by poison. Even should the view be perfectly accidental, or should there be only a suspicion that the proscribed articles have been seen, no mercy is shown; and it is said that fathers have been the executioners of their own daughters, and husbands of their wives, when such has been the case. I was of course anxious to purchase articles to which such curious customs belong, and spoke to the Tushaúa on the subject. He at length promised to sell them me on my return, stipulating that they were to be embarked at some distance from the village, that there might be no danger of their being seen by the women.

On the morning previous to that on which we were to leave, two more of our Indians who had received full payment on starting, were discovered to have left us. They had taken possession of a canoe, and absconded in the night; leaving me no remedy, but the chance of finding them in their houses on my return, and the still more remote chance of their having anything to pay me with.

The Indians here have but little characteristic distinction from those below. The women wear more beads around their necks and arms. The lower lip is often pierced, and two or three little strings of white beads inserted; but as the nations are so mixed by inter-marriages, this custom is probably derived from the Tucanos. Some of the women and children wore two garters, one above the ankle and one below the knee—swelling out the calf enormously, which they consider a very great beauty. I did not see here so many long tails of hair; most of the men having probably been to the Rio Negro with some trader, and thence worn their hair like Christians; or perhaps because the last Tushaúa was a "homen muito civilizado" (a very well-bred person).

After four days' delay, we at length started, with a comparatively small complement of Indians, but with some extra men to assist us in passing several caxoeiras, which occur near at hand. These are the "Piréwa" (Wound), "Uacorouá" (Goat-sucker), "Maniwára" (White Ant), "Matapí" (Fish-trap), "Amána" (Rain), "Tapíracúnga" (Tapir's head), "Tapíra eura" (Tapir's mouth), and "Jacaré" (Alligator). Three of these were very bad, the canoe having to be unloaded entirely, and pulled over the dry and uneven rocks. The last was the highest; the river rushing furiously about twenty feet down a rugged slope of rock. The loading and unloading of the canoe three or four times in the course of as many hours, is a great annoyance. Baskets of farinha and salt, of mandiocca-cakes [a starchy "cake" of manioc flour, which is from the root of the *Manihot*, or cassava plant] and pacovas [packet made of large leaves, usually banana leaves, for food], are strewn about. Panellas [pans] are often broken; and when there comes a shower of rain, everything has to be heaped together in a hurry,—palm-leaves cut, and the more perishable articles covered; but boxes, *rédés* [hammocks], and numerous other articles are sure to be wetted, rendering us very uncomfortable when again hastily tumbled into the overcrowded canoe. If I had birds or insects out drying, they were sure to be overturned, or blown by the wind, or wetted by the rain, and the same fate was shared by my note-books and papers. Articles in boxes, unless packed tight, were shaken and rumpled by not being carried evenly; so that it was an excellent lesson in patience, to bear all with philosophical serenity. We had passed all these falls by midday; and at night slept on a rock, where there was a small rapid and a house without inhabitants.

On the 8th we had tolerably quiet water, with only two small rapids, the "Taiéna" (Child), and "Paroquet" caxoeiras. On the 9th, in the morning, we reached the "Pacu" fall, and then had a quiet stream, though full of rocks, till the afternoon, when we passed the "Macucú" (a tree), "Ananás" (Pineapple), and "Uacú" (a fruit) caxoeiras; all very bad and difficult ones. We had left Carurú with very little farinha, as none was to be had there, and we had seen no inhabited sitios where any could be purchased; so our Indians were now on short allowance of "beijú" [bread made from manioc (cassava) flour], which they had brought with them. Of a passing Indian I bought a basket of Ocokí, and some fish. The Ocokí is a large pear-shaped fruit, with a hard thick outer skin of almost a woody texture, then a small quantity of very sweet pulpy matter, and within a large black oval stone. The pulp is very luscious, but is so acrid as to make the mouth and throat sore, if more than two or three are eaten. When, however, the juice is boiled it loses this property; and when made into mingau [a thick drink] with tapioca, is exceedingly palatable and very highly esteemed in the Upper Rio Negro, where it is abundant. It takes at least a peck of fruit to give one small panella of mingau.

On the next day, the 10th, in the afternoon, the Indians all suddenly sprang like otters into the water, swam to the shore, and disappeared in the forest. "Ocokí," was the answer to my inquiries as to the cause of their sudden disappearance; and I soon found they had discovered an ocokí-tree, and were loading themselves with the fruit to satisfy the cravings of hunger, for an Indian's throat and mouth seem invulnerable to all those scarifying substances which act upon civilised man. The tree is one of the loftiest in the forest, but the fruit falls as soon as ripe, and its hard woody coating preserves it from injury. Baskets, shirts, trousers, etc., were soon filled with the fruit and emptied into the canoe; and I made each of the Indians bring a small basketful for me; so that we had "mingau de ocokí" for three succeeding mornings.

The rocks from Carurú often present a scoriaceous appearance, as if the granite had been remelted. Sometimes they are a mass of burnt fragments, sometimes a honeycombed rock with a shining surface. In some places there are enclosed fragments of a finer-grained rock, apparently sandstone, and numerous veins and dykes, which often cross each other in three or four sets. The rocks are, in many places, so broken and cleft vertically, as to appear stratified and thrown up on end. The rounded form and concentric arrangement, observed in the Rio Negro, is here also constantly met with. The interstices of the rounded and angular masses of rock are often filled with a curious volcanic substance, which outwardly resembles pitch, but consists of scoriæ, sand, clays, etc., variously cemented together.

On the 10th we passed the "Tapioca," "Tucáno" (Toucan), "Tucunaré" (a fish), "Uaracú piními" (a fish), and "Tyeassú" (Pig) caxoeiras. The first was very bad, and both difficult and dangerous to pass; it consisted of many distinct falls among huge masses of rock. At one place the canoe remained stuck fast, amidst foaming waters, on the very edge of a fall, for nearly an hour; all the efforts of the Indians could not move it forward. They heaved it over from one side to the other, but with no effect; till I began to despair of getting out of the difficulty before night. At last the canoe suddenly moved on, with apparently not so much force as had been before applied to it; but my Indians, being of several nations, did not understand any common language, and it was impossible to get them to act in concert, or obey any leader. It was probably some chance combination of forces, that at last extricated us from our unpleasant situation. At this fall, on the rocks, were very numerous figures, or picture-writings, and I stopped to make drawings of them; of which I had by this time a rather extensive collection.

The next three falls were small rapids; but the last, which we reached late in the evening, was fearful. The river makes a sudden bend, and is confined in a very narrow channel, which is one confused mass of rocks of every size and shape, piled on one another, and heaped up in the greatest possible confusion. Every stone which rises above high-water mark is covered with vegetation; and among the whole the river rushes and foams, so as to make the task of pilot one of no ordinary difficulty. Just as it was getting dark, we passed out of these gloomy narrows into a wider and more cheerful part of the river, and stayed at a rock to sup and sleep.

On the 11th, early, we reached Uarucapurí, where are a village and several maloccas [see Figure 9]. The first which we entered was inhabited by people of the Cobeu nation. There were about a dozen handsome men, all clean-limbed and well painted, with armlets and necklaces of white beads, and with the ears plugged with a piece of wood the size of a common bottle-cork, to the end of which was glued a piece of porcelain presenting a white shining surface. We agreed with these men to help to pass our canoe up the falls, and then proceeded on our walk through the village. My old friend Senhor Chagas was here, and with him I breakfasted off a fine pirahíba [a very large Amazonian fish], which his men had caught that morning, and which was the first I had eaten since my illness.

With some difficulty I succeeded in buying two or three baskets of farinha; and being anxious to get to my journey's end, which was now near at hand, about midday we proceeded. Our pilot and his son left us, and we had now only six paddles; but four or five additional men came with us to

Fig. 9. Scene on the Rio Negro. The bridge crosses a rocky *caxoeira* (rapid), and the scene depicts a *malocca,* or village house, in addition to two smaller houses. From *A Narrative of Travels on the Amazon and Rio Negro,* 1853, facing p. 232.

pass the remaining caxoeiras, which were near. Close to the village we passed the "Cururú" (a toad), and "Murucututú" (an owl) falls, both rather bad; and, soon after, arrived at the "Uacoroúa" (Goatsucker), the last great fall on the river below the "Juruparí," which is many days further up. Here the river is precipitated over a nearly vertical rock, about ten feet high, and much broken in places. The canoe had to be entirely unloaded, and then pulled up over the rocks on the margin of the fall, a matter of considerable difficulty. To add to our discomfort, a shower of rain came on while the canoe was passing; and the Indians, as usual, having scattered the cargo about in great confusion, it had to be huddled together and covered with mats and palm-leaves, till the shower, which was luckily a short one, passed over. Loading again and proceeding onwards, we passed three small rapids, the "Tatu" (Armadillo), "Ocokí" (a fruit), and "Pirantérá" (a fish) caxoeiras; and our additional Indians here left us, with their payment of fish-hooks and arrow-heads, as we now had only smooth water before us. In the afternoon we passed a malocca, where one of the Indians wished to land to see his friends; and as we did not stay, at night he took his departure, and we saw no more of him.

Early the next morning we reached Mucúra, where two young Brazilians, whom I had met with below, were residing, trading for salsa. I was now in the country of the painted turtle and the white umbrella-bird, and I determined to make a stay of at least a fortnight, to try and obtain these much-desired rarities.

Messrs. Nicoláu and Bellarmine were both out, and their little palm-leaf huts were evidently quite inadequate to my accommodation. The only other house was a small Indian malocca, also made entirely of "palha" [straw]; and I agreed with the owner to let me have half of it, giving him a small knife and mirror in payment, with which he was well contented. We accordingly cleared and swept out our part of the house, unloaded and arranged our things, and I then sent my guardas to a malocca, in which there were said to be plenty of Indians, to see if they had any farinha or pacovas to dispose of; and also to let them know that I would purchase birds, or fish, or any other animals they could obtain for me. The men were all out; but the same afternoon they came in great force to see the "Branco," and make an attack on my fish-hooks and beads, bringing me fish, pacovas, farinha, and mandiocca-cake, for all of which one of these two articles was asked in exchange.

I was now settled at the limit of my expedition, for I could not think of going a week further up only to see Juruparf caxoeira,—wasting the little time I had to rest, before again descending. We had made a favourable voyage, without any serious accident, up a river perhaps unsurpassed for the difficulties and dangers of its navigation. We had passed fifty caxoeiras, great and small; some mere rapids, others furious cataracts, and some nearly perpendicular falls. About twenty were rapids, up which, by the help of a long sipó attached to the canoe, instead of a rope, we were pulled without much difficulty. About eighteen were very bad and dangerous, requiring the canoe to be partially unloaded where practicable, and all the exertions of my Indians, often with additional assistance, to pass; and twelve were so high and furious as to require the canoe to be entirely unloaded, and either pulled over the dry and often very precipitous rocks, or with almost equal difficulty up the margin of the fall. At Carurú, as I have said, four-and-twenty men were scarcely able to pull my empty canoe over the rock, though plentifully strewn with branches and bushes, to smooth the asperities which would otherwise much damage the bottom: this was the reason why I purchased the Tushaúa's smaller obá, to proceed; and it was well I did, or I might otherwise have had to return without ever reaching the locality I had at length attained.

The next day, the 13th, I was employed drawing some new fish brought me the preceding evening. My hunters went out and brought me nothing but a common hawk. In the afternoon, the father and brother of the Indian I had found in the house, arrived, and with their wives and families; so now, with my six Indians and two hunters, we were pretty full; some of them, however, slept in a shed, and we were as comfortably accommodated as could be expected. The wives of the father and two sons were perfectly naked, and were, moreover, apparently quite unconscious of the fact. The old woman possessed a "saía," or petticoat, which she sometimes put on, and seemed then almost as much ashamed of herself as civilised people would be if they took theirs off. So powerful is the effect of education and habit!

Having been told by Senhor Chagas that there was an excellent hunter in the Codiarí, a river which enters from the north a short distance above Mucúra, I sent Philippe, one of my guardas, to try and engage him, and also to buy all the living birds and animals he could meet with. The following day he returned, bringing with him one "Macaco barrigudo" (*Lagothrix Humboldtii*), and a couple of parrots. On most days I had a new fish or two to figure, but birds and insects were very scarce. This day Senhor Nicoláu returned. On my first arrival I had been told that he had a "tataruga pintata" (painted turtle) for me, but that he would give it me himself on his arrival; so I did not meddle with it, though my Indians saw it in a "corrál," in a small stream near the house. On arriving, he sent to fetch it, but found it had escaped, though it had been seen in its cage on the preceding day. I thus lost perhaps my only chance of obtaining a much-desired and probably undescribed river turtle, as the time of egg-laying was past, and they had now retired into the lakes, and become very scarce and difficult to be met with.

As my Indians were here doing nothing, I sent three of them with Sebastião up the Codiarí, with beads, hooks, mirrors, etc., to buy monkeys, parrots, or whatever else they could meet with, as well as some farinha, which I did not wish to be in want of again. I sent them with instructions to go for five or six days, in order to reach the last sitio, and purchase all that was to be had. In two days, however, they returned, having been no further than Philippe had gone, Sebastião saying that his companions would not go on. He brought me some parrots and small birds, bows, bird-skins, and more farinha than my canoe would carry, all purchased very dearly, judging by the remnant of articles brought back.

Being now in a part of the country that no European traveller had ever before visited, I exceedingly regretted my want of instruments to determine the latitude, longitude, and height above the sea. The two last I had no means whatever of ascertaining, having broken my boiling-point thermo-meter, and lost my smaller one, without having been able to replace either. I once thought of sealing up a flask of air, by accurately weighing which on my return, the density of air at that particular time would be obtained, and the height at which a barometer would have stood might be deduced. But, besides that this would only give a result equal to that of a single barome-ter observation, there were insuperable difficulties in the way of sealing up the bottle, for whether sealing-wax or pitch were used, or even should the bottle be hermetically sealed, heat must be applied, and at the moment of application would, of course, rarefy the air within the bottle, and so pro-duce in such a delicate operation very erroneous results. My observations, however, on the heights of the falls we passed, would give their sum as about two hundred and fifty feet; now if we add fifty for the fall of the river between them, we shall obtain three hundred feet, as the probable height of the point I reached above the mouth of the river; and, as I have every reason to believe that that is not five hundred feet above the sea, we shall obtain eight hundred feet as the probable limit of the height of the river above the sea-level, at the point I reached. Nothing, however, can accurately determine this fact, but a series of barometer or "boiling point" observa-tions; and to determine this height above the next great gall, and ascertain the true course and sources of this little-known but interesting and impor-tant river, would be an object worth the danger and expense of the voyage.

There is said to be a week's smooth water above this place, to the Juru-parí caxoeira, which is higher than any below it; and above this no other fall has been found, though traders have been ten or fifteen days up. They say the river still keeps as wide or wider than below,—that the water is as "white," or muddy, as that of the Solimões,—that many trees, birds, and fish peculiar to the Solimões are there found,—that the Indians have Span-ish knives, ponchos, and coins,—and relate that, higher up, there are extensive "campos," with cattle, and men on horseback. All these interesting particulars seem to show that the river has its sources in the great plains which extend to the base of the Andes, somewhat near where the sources of the Guaviare are placed in most maps; but the latter river, from all the information I can obtain, is much smaller, and has a much shorter course. Having only a pocket surveying sextant, without any means of viewing two objects much differing in brilliancy, I endeavoured to obtain the latitude as accurately as I could, first by means of the zenith-distance at noon, obtained by a plumb-line and image of the sun, formed by a lens of about fifteen inches focus; and afterwards, by the meridian altitude of a star, obtained on a calm night, by reflection in a cuya of water. I took much care to ensure an accurate result, and have every reason to believe that the mean of the two observations will not be more than two or three minutes from the truth.

My expectations of finding rare and handsome birds here were quite disappointed. My hunter and Senhor Nicoláu killed a few umbrella-birds of the Rio Negro species; but of the white bird such contradictory statements were given,—many knowing nothing whatever about it, others saying that it was sometimes, but very rarely seen,—that I am inclined to think it is a mere white variety, such as occurs at times with our blackbirds and starlings at home, and as are sometimes found among the curassow-birds and agoutis. Another bird, which I had been long searching for, the "anambé de catinga," a species of *Cyanurus* [a genus of birds found only in Africa and Indonesia. It is possible that Wallace was referring to Tinamous, an order of neotropical birds some of which occur in the region he visited.] was here shot; and before leaving, I obtained four or five specimens of it, and as many of the commoner black-headed species. One or two small birds, new to me, were also obtained; and these, with two or three scarce butterflies, and about a dozen new species of fish, composed my natural-history collections in this remote and unvisited district. This was entirely owing, however, to my unfortunate and unforeseen illness, for birds in great variety had been very abundant, but the time of the fruit was now over; fish and turtles, too, were in extraordinary plenty at the commencement of the fall of the river, two months back; and during that period, constituting the short summer in these districts, while I lay half dead at São Joaquim, insects were doubtless more numerous.

But as there was now no remedy I made myself as contented as I could,

and endeavoured at least to complete my collection of the arms, implements, and ornaments of the natives. The Indians here were mostly "Co-
beus," and I obtained several of their peculiar ornaments and dresses, to
add to my collection. I also took advantage of the visit of a Tushaúa, or
chief, who well understood the Lingoa Geral, to obtain a vocabulary of their
language.

Just as I was about to leave on my voyage down, I received a note from
Senhor Chagas, requesting, in the name of Tenente Jesuino, the loan of my
canoe, to ascend higher up the river; which, as the time of his stay was very
uncertain, I was obliged to refuse. This Tenente, an ignorant half-breed,
was sent by the new Barra government to bring all the Tushaúas, or chiefs,
of the Uaupés and Isanna rivers to Barra, to receive diplomas and presents.
An Indian, sent by him, had arrived at Carurú caxoeira, and wished to buy
the obá of the Tushaúa, after I had paid for and got possession of it, and
even had the impudence to request me to give it back again, in order that
he might purchase or borrow it; and my refusal was, of course, quite sufficient seriously to offend the said Tenente.

On the 25th, having been just a fortnight at Mucúra, I left, much disappointed with regard to the collections I had made there. The same day I
reached Uarucapurí, whence I could not proceed without a pilot, as the
falls below are very dangerous. There was hardly a male in the village,
Messrs. Jesuino and Chagas having taken all with them up the river, to assist in an attack on an Indian tribe, the "Carapanás," where they hoped to
get a lot of women, boys, and children, to take as presents to Barra. There
was scarcely anything to be had to eat: fish were not to be caught, though
we sent our Indians out every day; and though fowls were abundant, their
owners were out, and those in charge of them would not sell them. At
length, after four days, I succeeded in persuading the son of the Tushaúa
to go with me as pilot to Jauarité, he not being able to resist the knives,
beads, and mirror, which I spread out before him.

I had collected scarcely anything in this place, but a single specimen of
the beautiful and rare topaz-throated hummer (*Trochilus pyra*) and a new
butterfly of the genus *Callithea*. I heard of the handsome bronze *Jacana* [a
tropical bird with extremely long toes and claws that allow it to walk on
floating plants] being found here, but my hunters searched for it in vain.

On the morning after we left, we saw a fine deer on a sandbank near us,
so I sent Manoel into the forest to get behind it, while we remained quietly

watching from the canoe. After walking about the beach a short time, it took to the water to cross the river, when we followed in pursuit; and, notwithstanding its turnings and doublings, soon came up,—when the poor animal was despatched by a blow on the head, and pulled into the canoe. The Indians then went briskly on, rejoicing in the certainty of a dinner for the next day or two, in which I heartily joined them. At Tapioca caxoeira we stayed two hours, to cook and salt the deer, and descended the fall without any accident.

On April 1st we passed a host of falls, shooting most of them amidst fearful waves and roaring breakers, and arrived safely at Carurú, where the Tushaúa gave us his house; for, having two canoes, we were obliged to wait to get more Indians. I was still too weak to go out into the forest; and, besides, had my live stock to attend to, which now consisted of four monkeys, about a dozen parrots, and six or eight small birds. It was a constant trouble to get food for them in sufficient variety, and to prevent them from escaping. Most of the birds are brought up without being confined, and if placed in a cage, attempt constantly to get out, and refuse food till they die; if, on the other hand, they are loose, they wander about to the Indians' houses, or into the forest, and are often lost. I here had two new toldas made to my canoes, but all attempts to hire men were fruitless. Fowls and fish were tolerably abundant, so we were better off then at Uarucapurí.

On the 4th, in the afternoon, Senhors Jesuino and Chagas arrived with a whole fleet of canoes, and upwards of twenty prisoners, all, but one, women and children. Seven men and one woman had been killed; the rest of the men escaped; but only one of the attacking party was killed. The man was kept bound, and the women and children well guarded, and every morning and evening they were all taken down to the river to bathe. At night there was abundance of caxirí and caxaça drunk in honour of the new-comers, and all the inhabitants assembled in the great house. I spoke to Jesuino about obtaining some Indians for me, which he promised to do. Next morning, however, his first act was to summon my pilot, and scold him for coming with me at all,—frightening the poor fellow so, that he immediately went off with his father down the river. Before he had left, however, having been told by my guardas what was going on, I applied to Jesuino about the matter, when he denied having said anything to the pilot, but refused to call him back, or make him fulfill his engagement with me. Soon after Jesuino left, having first sent five Indians to take me to Jauarité;

so I started immediately after him. The men, however, had had instructions to go with me only a short distance, and then leave me where I could not procure any more; and about noon, much to my surprise, they got into a little obá, and intimated their intention to return, saying that they had only been told to come so far. I had overtaken Jesuino at this place, and now appealed to him; but though the men would have immediately obeyed an order from him he refused to give it, telling me that he had put them in my canoe, and now I must arrange with them as well as I could. I accordingly told the Indians, that if they came on with me to Jauarité, I would pay them well, but that, if they left me at this place, they should not have a single fish-hook; but they knew very well what Senhor Jesuino wanted, so without another word they paddled off, leaving me to get on as I could. I had now only one man and one boy in each canoe, to pass rapids which required six or eight good paddles to shoot with safety; but staying here was useless, so we went on,—drifting down the stream after Senhor Jesuino, who, no doubt, rejoiced in the idea that I should probably lose my canoes, if not my life, in the caxoeiras, and thought himself well revenged on the stranger who had dared to buy the canoe he had wanted to purchase.

In the afternoon we passed a caxoeira with considerable danger, and then, luckily, persuaded some Indians at a sitio to come with us to Jauarité. In the afternoon I stayed at several houses, purchasing fowls, parrots, bows and arrows and feathers; and at one of them I found my runaway pilot, and made him give me two baskets of farinha, instead of the payment he had received for the voyage from Carurú to Jauarité. At the last caxoeira, close to Jauarité, we were very near losing our canoe, which was let down by a rope, I remaining in it; but just in passing, it got twisted broadside, and the water rushing up from the bottom, had the curious effect of pushing it up against the fall, where it remained a considerable time completely on one side, and appearing as if every minute it would turn over. However, at last it was got out, and we reached the village, much to the surprise of Senhor Jesuino, who had arrived there but a few hours before us. My friend Senhor Augustinho, of São Jeronymo, was also there, and I spent the evening pleasantly with them.

I found that we differed in our calculations of the date, there being a day's difference in our reckonings of the day of the week and the day of the month. As I had been three months up the river, it was to be supposed I was wrong; yet as I had kept a regular diary all the voyage, I could not at all make out how I had erred. This, however, is a common thing in these remote districts. When two parties meet, one going up and the other coming down the river, the first inquiry of the latter, after the usual compliments, is, "What day is it with you?" and it not unfrequently happens, that there are three parties present, all of whom make it different days; and then there is a comparison of authorities, and a determination of past Saints' days, in order to settle the correction of the disputed calendar. When at Caturú caxoeira, we had found that Messrs. Jesuino and Chagas differed from us on this important particular; but as they had been some time out, we thought they might have erred as well as ourselves. Now, however, that Senhor Augustinho, who had recently come from São Gabriel, whence he had brought the correct date, agreed with them, there was no withstanding such authority. A minute examination of my diary was made, and it was then found that on our first stay at Carurú we had reckoned our delay there as five days instead of six. The Indians generally keep accounts of the time very accurately on a voyage, by cutting notches on a stick, as boys do at school on the approach of the holidays. In our case, however, even they were most of them wrong, for some of them agreed with me, while others made a day in advance, and others again a day behind us, so that we got completely confused. Sometimes the traders residing at the Indian villages pass many months, without seeing a person from any civilised part, and get two or three days out in their reckonings. Even in more populous places, where all the inhabitants depend on the priest or the commandante, errors have been made, and Sundays and Saints' days have been desecrated, while Mondays and common days have been observed in their place, much to the horror of all good Catholics.

The next morning I took a turn round the village,—bought some paroquets and parrots, and some feather ornaments and small pots, of the Tushaúa; and then, having nothing to keep me at Jauarité, and having vainly endeavoured to get some Indians to go with me, I left for São Jeronymo. On arriving at the first great fall of Pinupinú, we found only one Indian, and were obliged to send to the village for more. That afternoon they did not choose to come, and we lost a beautiful day. The next morning, as was to be expected, commenced a soaking rain; but as the Indians arrived we went on, and about noon, the rain clearing off a little, we passed the fall of Panoré, and arrived safely at the village of São Jeronymo. Here we disembarked, and unloaded our canoes, taking possession of the doorless "casa da nação [literally, house of the nation, a "public" house (not a pub)], and made up our minds to remain quietly till we should get men to go down the river.

The same afternoon Jesuino arrived, and the next morning left,—kindly inquiring when I intended to proceed, and saying, he had spoken with the Tushaúa to get me Indians. In two days, however, the Tushaúa also left for Barra, without giving me a single Indian, notwithstanding the promises and threats I had alternately employed.

The two Indians who had remained with me now left, and the two boys who had come from São Joaquim ran away, leaving me along in my glory, with my two "guardas" and two canoes. In vain I showed my axes, knives, beads, mirrors, and cloth, to every passing Indian; not one could be induced to go with me, and I might probably have remained prisoner there for months, had not Senhor Victorino, the "Juiz de Paz," arrived, and also Bernado, my old pilot, who had left me at Jauarité, and had now been down to São Joaquim. Between them, after a delay of several more days, some Indians were persuaded to receive payment to go with me as far as Castanheiro, where I hoped to get Capitão Ricardo to order them on to Barra.

~ ~

Sinking of the Helen

Brig, *Jordeson*, N. Lat. 49° 30' W. Long. 20°.
Sunday, September 19, 1852.

MY DEAR FRIEND,

Having now some prospect of being home in a week or ten days, I will commence giving you an account of the peculiar circumstances which have already kept me at sea seventy days on a voyage which took us only twenty-nine days on our passage out. I hope you have received the letter sent you from Para, dated July 9 or 10, in which I informed you that I had taken my passage in a vessel bound for London, which was to sail in a few days. On Monday, July 12, I went on board with all my cargo, and some articles purchased or collected on my way down, with the remnant (about twenty) of my

Letter from Wallace to Richard Spruce, *My Life*, 1:303–14.

live stock.* After being at sea about a week I had a slight attack of fever, and at first thought I had got the yellow fever after all. However, a little calomel set me right in a few days, but I remained rather weak, and spent most of my time reading in the cabin, which was very comfortable. On Friday, August 6, we were in N. Lat. 30° 30', W. Long. 52°, when, about nine in the morning, just after breakfast, Captain Turner, who was half-owner of the vessel, came into the cabin, and said, "I'm afraid the ship's on fire. Come and see what you think of it." Going on deck I found a thick smoke coming out of the forecastle, which we both thought more like the steam from heating vegetable matter than the smoke from a fire. The fore hatchway was immediately opened to try and ascertain the origin of the smoke, and a quantity of cargo was thrown out, but the smoke continuing without any perceptible increase, we went to the after hatchway, and after throwing out a quantity of piassaba [commercially important cordage for rope made from the fibers of the *Leopoldina piassaba* palm (Figure 10). Wallace demarcated the limited range of this palm and described it scientifically for the first time.], with which the upper part of the hold was filled, the smoke became so dense that the men could not stay in it. Most of them were then set to work throwing in buckets of water, and the rest proceeded to the cabin and opened the lazaretto or store-place beneath its floor, and found smoke issuing from the bulkhead separating it from the hold, which extended half-way under the fore part of the cabin. Attempts were then made to break down this bulkhead, but it resisted all efforts, the smoke being so suffocating as to prevent anyone stopping in it more than a minute at a time. A hole was then cut in the cabin floor, and while the carpenter was doing this, the rest of the crew were employed getting out the boats, the captain looked after his chronometer, sextant, books, charts, and compasses, and I got up a small tin box containing a few shirts, and put in it my drawings of fishes and palms, which were luckily at hand; also my watch and a purse with a few sovereigns. Most of my clothes were scattered about the cabin, and in the dense suffocating smoke it was impossible to

* These consisted of numerous parrots and parrakeets, and several uncommon monkeys, a forest wild-dog, etc.

Leopoldinia piassaba.

Fig. 10. Wallace's drawing of the palm, *Leopoldinia piassaba*, one of several new species of palm that Wallace described in his book *Palm Trees of the Amazon and Their Uses*. Fibers from the stem of this tree were (and still are) made into cordage and used commercially for rope, cable, and brooms. From the Linnean Society of London.

look about after them. There were two boats, the long-boat and the captain's gig, and it took a good deal of time to get the merest necessaries collected and put into them, and to lower them into the water. Two casks of biscuit and a cask of water were got in, a lot of raw pork and some ham, a few tins of preserved meats and vegetables, and some wine. Then there were corks to stop the holes in the boats, oars, masts, sails, and rudders to be looked up, spare spars, cordage, twine, canvas, needles, carpenter's tools, nails, etc. The crew brought up their bags of clothes, and all were bundled indiscriminately into the boats, which, having been so long in the sun, were very leaky and soon became half full of water, so that two men in each of them had to be constantly baling out the water with buckets. Blankets, rugs, pillows, and clothes were all soaked, and the boats seemed overloaded, though there was really very little weight in them. All being now prepared, the crew were again employed pouring water in the cabin and hatchway.

The cargo of the ship consisted of rubber, cocoa, anatto, balsam-capivi, and piassaba. The balsam was in small casks, twenty stowed in sand, and twenty small kegs in rice-chaff, immediately beneath the cabin floor, where the fire seemed to be. For some time we had heard this bubbling and hissing as if boiling furiously, the heat in the cabin was very great, flame soon broke into the berths and through the cabin floor, and in a few minutes more blazed up through the skylight on deck. All hands were at once ordered into the boats, which were astern of the ship. It was now about twelve o'clock, only three hours from the time the smoke was first discovered. I had to let myself down into the boat by a rope, and being rather weak it slipped through my hands and took the skin off all my fingers, and finding the boat still half full of water I set to baling, which made my hands smart very painfully. We lay near the ship all the afternoon, watching the progress of the flames, which soon covered the hinder part of the vessel and rushed up the shrouds and sails in a most magnificent conflagration. Soon afterwards, by the rolling of the ship, the masts broke off and fell overboard, the decks soon burnt away, the ironwork at the sides became red-hot, and last of all the bowsprit, being burnt at the base, fell also. No one had thought of being hungry till darkness came on, when we had a meal of biscuit and raw ham, and then disposed ourselves as well as we could for the night, which, you may be sure, was by no means a pleasant one. Our boats continued very leaky, and we could not cease an instant from baling; there was a considerable swell, though the day had been remarkably fine, and there were constantly floating around us pieces of the burnt wreck, masts, etc., which might have stove in our boats had we not kept a constant look-out to keep clear of them. We remained near the ship all night in order that we might have the benefit of its flames attracting any vessel that might pass within sight of it.

It now presented a magnificent and awful sight as it rolled over, looking like a whole caldron of fire, the whole cargo of rubber, etc., forming a liquid burning mass at the bottom. In the morning our little masts and sails were got up, and we bade adieu to the *Helen*, now burnt down to the water's edge, and proceeded with a light east wind towards the Bermudas, the nearest land, but which were more than seven hundred miles from us. As we were nearly in the track of West Indian vessels, we expected to fall in with some ship in a few days.

I cannot attempt to describe my feelings and thoughts during these events. I was surprised to find myself very cool and collected. I hardly thought it possible we should escape, and I remember thinking it almost foolish to save my watch and the little money I had at hand. However, after being in the boats some days I began to have more hope, and regretted not having saved some new shoes, cloth coat and trousers, hat, etc., which I might have done with a little trouble. My collections, however, were in the hold, and were irretrievably lost. And now I began to think that almost all the reward of my four years of privation and danger was lost. What I had hitherto sent home had little more than paid my expenses, and what I had with me in the *Helen* I estimated would have realized about £500. But even all this might have gone with little regret had not by far the richest part of my own private collection gone also. All my private collection of insects and birds since I left Para was with me, and comprised hundreds of new and beautiful species, which would have rendered (I had fondly hoped) my cabinet, as far as regards American species, one of the finest in Europe. Fancy your regrets had you lost all your Pyrenean mosses on your voyage home, or should you now lose all your

South American collection, and you will have some idea of what I suffer. But besides this, I have lost a number of sketches, drawings, notes, and observations on natural history, besides the three most interesting years of my journal, the whole of which, unlike any pecuniary loss, can never be replaced; so you will see that I have some need of philosophic resignation to bear my fate with patience and equanimity.

Day after day we continued in the boats. The winds changed, blowing dead from the point to which we wanted to go. We were scorched by the sun, my hands, nose, and ears being completely skinned, and were drenched continually by the seas or spray. We were therefore almost constantly wet, and had no comfort and little sleep at night. Our meals consisted of raw pork and biscuit, with a little preserved meat or carrots once a day, which was a great luxury, and a short allowance of water, which left us as thirsty as before directly after we had drunk it. Ten days and ten nights we spent in this manner. We were still two hundred miles from Bermuda, when in the afternoon a vessel was seen, and by eight in the evening we were on board her, much rejoiced to have escaped a death on the wide ocean, whence none would have come to tell the tale. The ship was the *Jordeson,* bound for London, and proves to be one of the slowest old ships going. With a favourable wind and all sail set, she seldom does more than five knots, her average being two or three, so that we have had a most tedious time of it, and even now cannot calculate with any certainty as to when we shall arrive. Besides this, she was rather short of provisions, and as our arrival exactly doubled her drew, we were all obliged to be put on strict allowance of bread, meat, and water. A little ham and butter of the captain's were soon used up, and we have been now for some time on the poorest of fare. We have no suet, butter, or raisins with which to make "duff," or even molasses, and barely enough sugar to sweeten our tea or coffee, which we take with dry, coarse biscuit, and for dinner, beef or pork of the very worst quality I have ever eaten or even imagined to exist. This, repeated day after day without any variation, beats even Rio Negro fare, rough though it often was. About a week after we were picked up we spoke and boarded an outward bound ship, and got from her some biscuits, a few potatoes, and some salt cod, which were a great improvement, but did not last long. We have also occasionally caught some dolphin and a few fish resembling the acarrás of the Rio Negro; but for some time now we have seen none, so that I am looking forward to the "flesh-pots of Egypt" with as much pleasure as when we were luxuriating daily on farina and "fiel amigo."* While we were in the boats we had generally fine weather, though with a few days and nights squally and with a heavy sea, which made me often tremble for our safety, as we heeled over till the water poured in over the boat's side. We had almost despaired of seeing any vessel, our circle of vision being so limited; but we had great hopes of reaching Bermuda, though it is doubtful if we should have done so, the neighbourhood of those islands being noted for sudden squalls and hurricanes, and it was the time of the year when the hurricanes most frequently occur. Having never seen a great gale or storm at sea, I had some desire to witness the phenomenon, and have now been completely gratified. The first we had about a fortnight ago. In the morning there was a strong breeze and the barometer had fallen nearly half an inch during the night and continued sinking, so the captain commenced taking in sail, and while getting in the royals and studding-sails, the wind increased so as to split the mainsail, fore-top-sail, fore-trysail, and jib, and it was some hours before they could be got off her, and the main-topsail and fore-sail double reefed. We then went flying along, the whole ocean a mass of boiling foam, the crests of the waves being carried in spray over our decks. The sea did not get up immediately, but by night it was very rough, the ship plunging and rolling most fearfully, the sea pouring in a deluge over the top of her bulwarks, and sometimes up over the cabin skylight. The next morning the wind abated, but the ship, which is a very old one, took in a deal of water, and the pumps were kept going nearly the whole day to keep her dry. During this gale the wind went completely round the compass, and then settled nearly due east, where it pertinaciously continued for twelve days, keeping us tacking about, and making less than forty miles a day against it. Three days ago we had another gale,

---

* This was the name given by our kind host, Señor Henrique, at Barra, to dried pi-rarucú, meaning "faithful friend," always at hand when other food failed.

more severe than the former one—a regular equinoctial which lasted two entire days and nights, and split one of the newest and strongest sails on the ship. The rolling and plunging were fearful, the bowsprit going completely under water, and the ship being very heavily laden with mahogany, fustic, and other heavy woods from Cuba, strained and creaked tremendously, and leaked to that extent that the pumps were obliged to be kept constantly going, and their continued click-clack, click-clack all through the night was a most disagreeable and nervous sound. One day no fire could be made owing to the sea breaking continually into the galley, so we had to eat a biscuit for our dinner; and not a moment's rest was to be had, as we were obliged to be constantly holding on, whether standing, sitting, or lying, to prevent being pitched about by the violent plunges and lurches of the vessel. The gale, however, has now happily passed, and we have a fine breeze from the north-west, which is taking us along six or seven knots— quicker than we have ever gone yet. Among our other disagreeables here we have no fresh water to spare for washing, and as I only saved a couple of shirts, they are in a state of most uncomfortable dirtiness, but I console myself with the thoughts of a glorious warm bath when I get on shore.

October 1. Oh, glorious day! Here we are on shore at Deal [port in southeastern England, north of Dover], where the ship is at anchor. Such a dinner, with our two captains! Oh, beef-steaks and damson tart, a paradise for hungry sinners.

October 5. London. Here I am laid up with swelled ankles, my legs not being able to stand work after such a long rest in the ship. I cannot write now at any length—I have too much to think about. We had a narrow escape in the Channel. Many vessels were lost in a storm on the night of September 29, but we escaped. The old "Iron Duke" is dead [Duke of Wellington]. The Crystal Palace [a giant glass and iron exhibition hall in Hyde Park, London. It housed the Great Exhibition of the Works of Industry of All Nations of 1851. The palace was a new conception of building, unprecedented in the history of architecture.] is being pulled down, and is being rebuilt on a larger and improved plan by a company. Loddige's collection of plants has been bought entire to stock it, and they think by heating it in the centre to get a gradation of climates, so as to be able to have the plants of different countries, tropical or temperate, in one undivided building. This is Paxton's plan. [The Crystal Palace was designed by the architect Sir Joseph Paxton; it was taken down and rebuilt in 1852–54.]

How I begin to envy you in that glorious country where "the sun shines for ever unchangeably bright," where farina abounds, and of bananas and plantains there is no lack! Fifty times since I left Para have I vowed, if I once reached England, never to trust myself more on the ocean. But good resolutions soon fade, and I am already only doubtful whether the Andes or the Philippines are to be the scene of my next wanderings. However, for six months I am a fixture here in London, as I am determined to make up for lost time by enjoying myself as much as possible for awhile. I am fortunate in having about £200 insured by Mr. Stevens' foresight, so I must be contented, though it is very hard to have nothing to show of what I took so much pains to procure.

I trust you are well and successful. Kind remembrances to everybody, everywhere, and particularly to the respectable Senhor João de Lima of São Joachim.

Your very sincere friend,
Alfred R. Wallace

Some of the most alarming incidents, to a landsman, are not mentioned either in this letter or in my published "Narrative." The captain had given the only berths in the cabin to Captain Turner and myself, he sleeping on a sofa in fine weather, and on a mattress on the floor of the cabin when rough. On the worst night of the storm I saw him, to my surprise, bring down an axe and lay it beside him, and on asking what it was for, he replied, "To cut away the masts in case we capsize in the night." In the middle of the night a great sea smashed our skylight and poured in a deluge of water, soaking the poor captain, and then slushing from side to side with every roll of the ship. Now, I thought, our time is come; and I expected to see the captain rush up on deck with his axe. But he only swore a good deal, sought out a dry coat and blanket, and then lay down on the sofa as if nothing had happened. So I was a little reassured.

Not less alarming was the circumstance of the crew coming aft in a body to say that the forecastle was uninhabitable, as it was constantly wet, and several of them brought handfuls of wet rotten wood which they could pull out in many places. This happened soon after the first gale began; so the two captains and I went to look, and we saw sprays and squirts of water coming in at the joints in numerous places, soaking almost all the men's berths, while here and there we could see the places where they had pulled out rotten wood with their fingers. The captain then had the sail-room amidships cleared out for the men to sleep in for the rest of the voyage.

One day in the height of the storm, when we were being flooded with spray and enormous waves were coming up behind us, Captain Turner and I were sitting on the poop in the driest place we could find, and, as a bigger wave than usual rolled under us and dashed over our sides, he said quietly to me, "If we are pooped by one of those waves we shall go to the bottom"; then added, "We were not very safe in our two small boats, but I had rather be back in them where we were picked up than in this rotten old tub." It is, therefore, I think, quite evident that we *did* have a very narrow escape. Yet this unseaworthy old ship, which ought to have been condemned years before, had actually taken Government stores out to Halifax, had there been patched up, and sent to Cuba for a cargo of heavy timber, which we were bringing home.

I may here make a few remarks on the cause of the fire, which at the time was quite a mystery to us. We learnt afterwards that balsam-capivi is liable to spontaneous combustion by the constant motion on a voyage, and it is for that reason that it is always carried in small kegs and imbedded in damp sand in the lowest part of the hold. Captain Turner had never carried any before, and knew nothing of its properties, and when at the last moment another boat load of small kegs of balsam came with no sand to pack them in, he used rice-chaff which was at hand, and which he thought would do as well and this lot was stored under the cabin floor, where the flames first burst through and where the fire, no doubt, originated.

Captain Turner had evidently had no experience of fire in a ship's cargo, and took quite the wrong way in the attempt to deal with it. By opening the hatchways to pour in water he admitted an abundance of air, and this was what changed a smouldering heat into actual fire. If he had at once set all hands at work caulking up every crack through which smoke came out, making the hatchways also air-tight by nailing tarpaulins over them, no flame could have been produced, or could have spread far, and the heat due to the decomposition of the balsam would have been gradually diffused through the cargo, and in all probability have done no harm. A few years later a relative of mine returning home from Australia had a somewhat similar experience, in which the captain adopted this plan and saved the ship. When in the Indian Ocean some portion of the cargo was found to be on fire, by smoke coming out as in our case. But the captain immediately made all hatches and bulkheads air-tight; then had the boats got out and prepared for the worst, towing them astern; but he reached Mauritius in safety, and was there able to extinguish the fire and save the greater part of the cargo.

On the receipt of my letter Dr. Spruce, who was then, I think, somewhere on the Rio Negro or Uaupés, wrote to the "João de Lima," referred to by me (and usually mentioned in my "Travels" as Senhor L.), giving him a short account of my voyage home; and a few months later he received a reply from him. He was a Portuguese trader who had been many years resident on the upper Rio Negro, on whose boat I took a passage for my first voyage up the river, and with whom I lived a long time at Guia. I also went with him on my first voyage up the river Uaupés. He was a fairly educated man, and had an inexhaustible fund of anecdotes of his early life in Portugal, and would also relate many "old-time" stories, usually of the grossest kind, somewhat in the style of Rabelais, or of Chaucer's coarsest Canterbury tales. Old Jeronymo was a quiet old man, a half-bred Indian, or Mameluco as they were called, who lived with Senhor Lima as a humble dependent, assisting him in his business and making himself generally useful. It was these two who were with me during my terrible fever, and who one night gave me up as certain not to live till morning. Dr. Spruce gave me this letter, and as it mainly refers to me, I will here give a nearly literal translation of it.

San Joaquim, June 7, 1853.

Illustrissimo Senhor Ricardo Spruce:

I received your greatly esteemed favour dated the 26th April last, and was rejoiced to hear of your honour's health and all the news that you give me, and I was much grieved at the misfortunes which befell our good friend Alfredo! My dear Senhor Spruce, what labours he performed for mankind, and what trouble to lose all his work of four years; but yet his life is saved, and that is the most precious for a man! Do me the favour, when you write to Senhor Alfredo, to give my kind remembrances. The mother of my children also begs you to give her remembrances to Senhor Alfredo, also tell him from me that if he eve comes to these parts again he will find that I shall be to him the same Lima as before, and give him more remembrances from the bottom of my heart, and also to yourself, from

Yours, with much affection and respect,
João Antonio de Lima.

N.B.—Old Jeronymo also asks you to remember him to Senhor Alfredo, and to tell him that he still has the shirt that Senhor Alfredo gave him, and that he is still living a poor wanderer with his friend Lima.

On reaching London in the condition described in my letter to Dr. Spruce, and my only clothing a suit of the thinnest calico, I was met by my kind friend and agent, Mr. Samuel Stevens, who took me first to the nearest ready-made clothes shop, where I got a warm suit, then to his own tailor, where I was measured for what clothes I required, and afterwards to a haberdasher's to get a small stock of other necessaries. Having at that time no relatives in London, his mother, with whom he lived in the south of London—I think in Kensington—had invited me to stay with her. Here I lived most comfortably for a week, enjoying the excellent food and delicacies Mrs. Stevens provided for me, which quickly restored me to my usual health and vigour.

Since I left home, and after my brother John had gone to California in 1849, my sister had married Mr. Thomas Sims, the elder son of my former host at Neath. Mr. Sims had taught himself the then rapidly advancing art of photography, and as my sister could draw very nicely in water-colours, they had gone to live at Weston-super-Mare, and established a small photographic business. As I wished to be with my sister and mother during my stay in England, I took a house then vacant in Upper Albany Street (No. 44), where there was then no photographer, so that we might all live together. While it was getting ready I took lodgings next door, as the situation was convenient, being close to the Regent's Park and Zoological Gardens, and also near the Society's offices in Hanover Square, and with easy access to Mr. Stevens's office close to the old British Museum. At Christmas we were all comfortably settled, and I was able to begin the work which I had determined to do before again leaving England. . . .

~⭐ ⭐~

On the Monkeys of the Amazon

The great valley of the Amazon is rich in species of Monkeys, and during my residence there I had many opportunities of becoming acquainted with their habits and distribution. The few observations I have to make will apply principally to the latter particular. I have myself seen twenty-one species; seven with prehensile and fourteen with non-prehensile tails, as shown in the following list:—

3 Howlers, viz.—*Mycetes ursinus, M. caraya?* and *M. beelzebub;*

1 Spider Monkey,—*Ateles paniscus;*

1 Big-bellied Monkey (*Barrigudo* of the Brazilians),—*Lagothrix humboldtii;*

2 Sapajou,—*Cebus gracilis* (Spix) and *C. apella?;*

4 Short-tailed Monkeys,—*Brachyurus couxiu, B. ouakari* (Spix), *B. rubicundus* (? *Calvus,* B. M.), and a new species;

2 Sloth Monkeys,—*Pithecia irrorata* and an undescribed species;

3 Squirrel Monkeys,—*Callithrix sciureus, C. personatus* and *C. torquatus;*

2 Nocturnal Monkeys,—*Nyctipithecus trivirgatus* and *N. felinus;* and

3 Marmoset Monkeys,—*Jacchus bicolor, J. tamarin* and a new species.

The Howling Monkeys are generally abundant; the different species, however, are found in separate localities; *Mycetes beelzebub* being apparently confined to the Lower Amazon, in the vicinity of Para; a black species, *M. caraya?*, to the Upper Amazon; and a red species, *M. ursinus,* to the Rio Negro and Upper Amazon. Much confusion seems to exist with regard to

Proceedings of the Zoological Society of London 20 (1852):107–10.

the species of Howlers, owing to the difference of colour in the sexes of some species. The red and the black species of the Amazon, however, are of the same colour in both sexes. The species of this genus are seminocturnal in their habits, uttering their cries late in the evening and before sunrise, and also on the approach of rain. Humboldt observes, that the tremendous noise they make can only be accounted for by the great numbers of individuals that unite in its production. My own observations, and the unanimous testimony of the Indians, prove this not to be the case. One individual only makes the howling, which is certainly of a remarkable depth and volume and curiously modulated; but on closely remarking the suddenness with which it ceases and again commences, it is evident that it is produced by one animal, which is generally a full-grown male. On dissecting the throat, much of our wonder at the noise ceases; for besides the bony vessel formed by the expanded "os hyoides," there is a strong muscular apparatus which seems to act as a bellows in forcing a body of air through the reverberating bony cavity.

Of the genus *Ateles*, the four-fingered Spider Monkeys, one species is found only in the Guiana district, north of the Amazon and Rio Negro. Another, probably *Ateles ater*, inhabits the West Brazil district on the river Purus. These monkeys are slow in their motions, but make great use of their prehensile tail, by which they swing themselves from bough to bough; and I have been informed that two have been seen to join together by their hands and prehensile tails, to form a bridge for their young ones to pass over. The Indians also say, that this animal generally moves suspended beneath the boughs, not walking on them.

The next genus, *Lagothrix*, is a very interesting one, being quite unknown in Guiana and Eastern Brazil. The species I am acquainted with (*L. Humboldtii*) is found in the district south-west of the Rio Negro, towards the Andes, which I call the Ecuador district of the Amazon. They are remarkable for their thick woolly grey fur, their long prehensile tails, and very mild disposition. In the upper Amazon they are the species most frequently seen tame, and are great favourites, from their grave countenances, more resembling the human face than those of any other Monkeys, their quiet manners, and the great affection and docility they exhibit. I had three of them for several months before leaving Brazil, and they were on board with me at the time the ship was burnt, when, with their companions, they all perished.

The Sapajou Monkeys, forming the genus *Cebus*, appear to be more generally distributed, and the species have a wider range. They are also frequently domesticated, but offer a remarkable contrast to the species of the last genus, in their constant activity and restlessness, and they have the character of being the most mischievous monkeys in the country.

Each species of the genus *Brachyurus* appears to be confined to a particular district. The *B. couxiu* is a native of Guiana, and does not pass the Rio Negro on the west, or the Amazon on the south. The *B. ouakari* is found on the Upper Rio Negro; the *B. rubicundus* on the Upper Amazon, called the Solimoes; and another species, apparently undescribed, is found on the lower part of the same river.

The Sloth Monkeys, forming the genus *Pithecia*, have an extensive range as regards the genus, but the separate species seem each confined in a limited space. Of the two species inhabiting the Amazon district, one, the *P. irrorata*, is found on the south bank of the Upper Amazon; and another, apparently undescribed and rendered remarkable by a bright red beard round the face and under the chin, occurs only to the south-west of the Rio Negro.

Of the little Squirrel Monkeys, one, the *Callithrix sciureus*, a specimen of which is now in the Society's Gardens, has an extensive range, being found on both banks of the Amazon and Rio Negro. The *C. torquatus*, a white-collared species, is found only on the Upper Rio Negro, and the *C. personatus* on the Upper Amazon.

Of the curious Nocturnal Monkeys forming the genus *Nyctipithecus* there are two species in this district; one, which appears to be the *N. trivirgatus* of Humboldt, is found in the district of Ecuador, west of the Upper Rio Negro; the other, closely allied, probably the *N. felinus*, on the Upper Amazon. Their large eyes, cat-like faces, soft woolly hair and nocturnal habits render them a very interesting group. They are called "devil monkeys" by the Indians, and are said to sleep during the day and to roam about only at night. I have had specimens of them alive, but they are very delicate and soon die.

Of the Marmozet Monkeys there are three species, though none of them have the characteristic tufts of hair on the head. Each species seems to be confined to a very limited tract of country. The *Jacchus tamarin* is found only in the district of Para, where it is abundant. The *J. bicolor*, a pretty grey and white species, I have only seen on the Guiana side of the Rio Negro near the city of Barra. Another species entirely black, with the face of bare white skin, inhabits the district of the Upper Rio Negro. It appears to be quite new.

The last three genera appear to be to a great extent insectivorous, and I am inclined to think they also devour small birds and mammalia. At least those I have had alive would attempt to pull into their cages any of my small birds which passed near. The little black *Jacchus* last mentioned was particularly savage. He once seized a large parrot by the neck, pulled him into his cage, and bit out a large piece from his bill, and would probably have destroyed it, had I not opportunely come to the rescue. Two other small birds which approached too near his cage he seized and completely devoured.

I will now make a few remarks on the geographical distribution of these animals.

In the various works on natural history and in our museums, we have generally but the vaguest statements of locality. S. America, Brazil, Guiana, Peru, are among the most common; and if we have "River Amazon" or "Quito" attached to a specimen, we may think ourselves fortunate to get anything so definite: though both are on the boundary of two distinct zoological districts, and we have nothing to tell us whether the one came from the north or south of the Amazon, or the other from the east or the west of the Andes. Owing to this uncertainty of locality, and the additional confusion created by mistaking allied species from distant countries, there is scarcely an animal whose exact geographical limits we can mark out on the map.

On this accurate determination of an animal's range many interesting questions depend. Are very closely allied species ever separated by a wide interval of country? What physical features determine the boundaries of species and of genera? Do the isothermal lines [lines drawn on a map connecting points having the same temperature at a given time or same mean temperature for a given period, such as a winter isotherm] ever accurately bound the range of species, or are they altogether independent of them? What are the circumstances which render certain rivers and certain mountain ranges the limits of numerous species, while others are not? None of these questions can be satisfactorily answered till we have the range of numerous species accurately determined.

During my residence in the Amazon district I took every opportunity of determining the limits of species, and I soon found that the Amazon, the Rio Negro and the Madeira formed the limits beyond which certain species never passed. The native hunters are perfectly acquainted with this fact, and always cross over the river when they want to procure particular animals, which are found even on the river's bank on one side, but never by any chance on the other. On approaching the sources of the rivers they cease to be a boundary, and most of the species are found on both sides of them. Thus several Guiana species come up to the Rio Negro and Amazon, but do not pass them; Brazilian species on the contrary reach but do not pass the Amazon to the north. Several Ecuador species from the east of the Andes reach down into the tongue of land between the Rio Negro and Upper Amazon, but pass neither of those rivers, and others from Peru are bounded on the north by the Upper Amazon, and on the east by the Madeira. Thus there are four districts, the Guiana, the Ecuador, the Peru and the Brazil districts, whose boundaries on one side are determined by the rivers I have mentioned.

In going up the Rio Negro the difference in the two sides of the river is very remarkable.

In the lower part of the river you will find on the north the *Jacchus bicolor* and the *Brachyurus couxiu,* and on the south the red-whiskered *Pithecia.* Higher up you will find on the north the *Ateles paniscus,* and on the south the new black *Jacchus* and the *Lagothrix Humboldtii.*

Spix, in his work on the monkeys of Brazil, frequently gives, "banks of the river Amazon" as a locality, not being aware apparently that the species found on one side very often do not occur on the other, though the fact is generally known to the natives. In these observations I have only referred to the monkeys, but the same phænomena occur both with birds and insects, as I have observed in many instances.

3

The Malay Archipelago

It is a sure sign of Wallace's passion for nature that within months of his disastrous return from Brazil in 1852 he was making plans for another journey to the tropics to collect animals. He seemed to know what he was after, and his publications from his years in the Malay Archipelago, 1854–62, tell us that he found it. The three papers he wrote between 1855 and 1858 outline his theory of organic evolution. The last of these, Wallace's now-famous paper of 1858, the second selection in this chapter, so startled Darwin that he abandoned his big book project and published an abstract of his theory, better known as *On the Origin of Species* (1859). The first reading depicts Wallace's search for the bird of paradise, one of the compelling aims of his journey.

Wallace departed for the Malay Archipelago (now Malaysia, part of New Guinea, and the islands that constitute Indonesia) in March 1854, well armed with books and supplies and accompanied by a young assistant.[1] See Figure 11 for a map of the region.

Wallace's contacts with Europeans living in the region formed the lattice of his journey. Through these connections, he borrowed, rented, or built the seventy or more houses he inhabited during eight years of fieldwork, and these to a large extent determined the locations of his daily

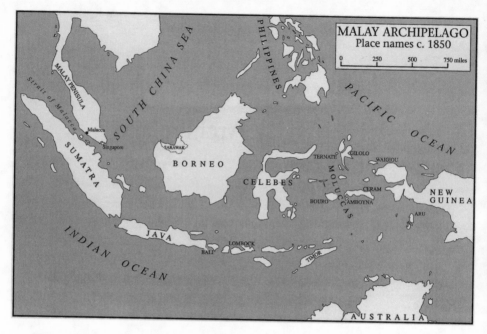

Fig. 11. Map of place names in the Malay Archipelago. Prepared by the University of Wisconsin Cartographic Laboratory.

collecting sites. His activities required the cooperation of other local peoples, both rulers and ordinary inhabitants, and these interactions were often mediated by local Europeans. One of Wallace's key contacts was Sir James Brooke, the celebrated British ruler of Sarawak (northwestern Borneo), who offered hospitality to his compatriot and provided introductions to other resident Europeans [in the region]. For nine months Wallace collected at a unique site that was being cleared by a British engineer and his Chinese laborers in preparation for building railroad tracks. The clearings and felled trees, surrounded by hundreds of miles of swamps and forest, provided the richest opportunity for collecting insects Wallace ever had as well as access to the isolated jungles inhabited by the orangutan.

The great manlike ape of Borneo, or Mias as it was called by the native Dyak people, had been studied in Europe from about a dozen skeletons, and a few live specimens were prominently but briefly displayed in London's Regent's Park. Observations of these great apes in their native habitat were virtually nonexistent; whatever Wallace could observe and collect of these animals would be valued by scientists and a public eager to know more about these peculiar creatures. Early in his eastern travels, Wallace sent back several accounts of hunting and observing orangutans in their native habitat in addition to selling the skins and bones of some fifteen specimens.² For three months he cared for an infant orangutan whose mother he had killed. Much to his dismay, the infant did not survive, but Wallace recorded its development in detail and with great affection. In one of the papers he discussed the difficulty of accounting for the large canine teeth of the males. He argued that living things are part of a larger, interconnected whole and that not every structure of every plant or animal has to be useful. Although Wallace had not yet formulated his theory of natural selection, he believed in common descent, and as early as 1856 we find him strongly suggesting that some traits may be best understood as part of a grander scheme.

Throughout his long journey, Wallace depended on native assistants to carry his supplies while he was traveling, to cook, to guard his belongings when he was out collecting, to carry nets and vials in the field, and sometimes to build temporary living quarters. He employed dozens of natives from the various islands, but one native deserves special notice.

Ali, a Malay servant Wallace employed in 1855, was a most willing and able assistant. In presenting Ali's photograph in his autobiography, Wallace described him as the faithful companion of nearly all of his journeys in the East and the best native servant he ever had (Figure 12). Ali was Wallace's head man, accompanying him on his daily collecting trips; together they walked, netted insects, shot birds, sat and ate in the field, and coordinated their evening chores and meals. Although Wallace often had one or more other servants as well, it was Ali who became skilled at collecting, who trained other servants, whose calm seamanship could be counted on, whom Wallace remembered as having saved his life during bouts of malaria, who stayed the course until the end, and who took on and kept the name "Wallace."³ Ali did more than serve as Wallace's eyes, ears, and hands in his collecting and as a domestic servant; he was also a teacher in the native language and ways, and in a very real sense, a friend and companion. It is not surprising that when Wallace departed for England, he left Ali a rich man, giving him cash, two guns, ammunition, and many stores and tools.

One of Wallace's most cherished goals in traveling to the Malay Archi-

Fig. 12. Wallace's Malay servant, Ali. "He here, for the first time, adopted European clothes, which did not suit him nearly so well as his native dress, and thus clad a friend took a very good photograph of him. I therefore now present his likeness to my readers as that of the best native servant I ever had, and the faithful companion of almost all my journeyings among the islands of the far East." From *My Life*, 1:383; photograph facing p. 382.

pelago was to collect the much sought-after but rare birds of paradise, long considered the most beautiful birds on earth. They stirred strong emotions, not least among them curiosity and greed. They are spectacular creatures, as unusual as they are exquisite and showy. The courtship plumage of the adult males accounts for most of the extraordinary feathers, with courtship dances and movements adding to their startling appearance (see Figures 13, 16, and 20). Chinese and Arab merchants traded for feathers of the birds of paradise long before Wallace came on the scene. Subsequent exploration and exploitation of the region by the Portuguese, Spanish, Dutch, and British added to the mythological status and high value accorded to the source of these exotic plumes. Wallace longed to see the birds of paradise in nature, both for the thrill their beauty excited and to add to Europeans' limited knowledge of them. He knew they would be hard to locate, but he did not appreciate quite how difficult it would turn out to be.

Unlike other islands that Wallace visited, the home of the birds of paradise had not been settled by Europeans and offered little in the way of colonial amenities. The Aru Islands, the eastern Moluccas, and New Guinea were obstacles to the naturalist for good reason. Food was scarce, the coastline and terrain were rugged and inhospitable, the swamps and forests nearly impassable, and the threat of serious illness and violent danger ever present. The absence of European settlement in the eastern end of the Malay Archipelago intensified Wallace's dependence on local peoples. Ali's role, as well as that of other servants, was nowhere more pronounced than in the journeys Wallace made in search of these birds.

Wallace made three separate trips expressly to collect birds of paradise: one trip to the Aru Islands in 1857, the next to mainland New Guinea in 1858, and the third in 1860 to Waigiou, the westernmost of the Molucca Islands (see Figure 11). On another journey to the island of Batchian late in 1858 Ali collected what Wallace recognized as a new species of bird of paradise (Figure 13). In the selection reproduced here Wallace narrates the first of these collecting trips. Wallace and his servants Ali, Baderoon, and Baso had spent two months based in the trading town of Dobbo on one of the Aru Islands (Figure 14). Disappointed by the collecting there and thwarted by bad weather and pirates, Wallace and his men finally sailed to Maykor, which is what Wallace called the mainland of Aru. This sometimes painful adventure led Wallace to the native habitat of both the great bird of paradise and the king bird of paradise.

When he returned to Celebes in July 1857 from his first major expedition to the Aru Islands, Wallace found seven months of accumulated mail; he also had nine thousand specimens from Aru to label, pack, and

Fig. 13. Wallace's standard wing (male and female), a new species of bird of paradise first found by Ali, recognized as a new form by Wallace, and later named at the British Museum as *Semioptera wallacei*. From *The Malay Archipelago*, facing p. 336; original wood engraving by J. G. Keulemans.

Fig. 14. Dobbo, a port and trading settlement in the Aru Islands. The engraving by Thomas Baines was made from a sketch by Wallace. From *The Malay Archipelago,* facing p. 476.

ship, an article on the natural history of Aru to write, and letters to answer, but he energetically completed his tasks. He responded to a letter from Charles Darwin, who had written that he agreed "to the truth of almost every word" of Wallace's 1855 "Law" paper (see the Introduction) and that after twenty years of study he was preparing for publication his work on how species and varieties differ. Only a scrap remains of Wallace's reply, in which he wrote that he had plans for a detailed proof of his theory on the succession of species.[4] Buoyed by his successful Aru collections and encouraged by Darwin's approval of his views, Wallace forged ahead in his thinking and specimen hunting. Early in 1858 he traveled from Celebes to the neighboring islands of Ternate and Gilolo. On Gilolo, he wrote "On the Tendency of Varieties to Depart Indefinitely from the Original Type," the paper that documents Wallace's independent formulation of the theory of evolution by natural selection.

In the spring of 1858, when Darwin received a letter from Wallace with a copy of this essay, his reaction was anything but subdued. He wrote to his mentor, the noted geologist Sir Charles Lyell, "Your words have come true with a vengeance that I should be forestalled. You said this when I explained to you here very briefly my views of 'Natural Selection' depending on the Struggle for existence. I never saw a more striking coincidence. if [*sic*] Wallace had my M.S. sketch written out in 1842 he could not have made a better short abstract! Even his terms now stand as Heads of my Chapters."[5] Struck by the glaring similarities, Darwin momentarily overlooked many of the differences between his and Wallace's theories. Paramount for Darwin was a powerful and almost violent sense that the two decades he had labored on his evolutionary theory were for naught: "So all my originality, whatever it may amount to, will be smashed."[6]

In the letter to Darwin which accompanied the essay, Wallace asked that it be forwarded to Sir Charles Lyell if Darwin found it sufficiently important, which he did.[7] Lyell then consulted with Joseph D. Hooker, noted botanist and confidant of Darwin, and they carefully arranged to present to the Linnean Society the following items: (1) a letter from Hooker and Lyell describing Darwin's and Wallace's independent formulation of the same theory; (2) extracts from an unpublished essay written in 1844 by Darwin in which he outlined his theory; (3) an abstract of a letter dated 1857 from Darwin to the American botanist Asa Gray in which Darwin described his views; and (4) the essay by Wallace. Thus Wallace's essay was published as the final part of a joint paper with Darwin as presented by Lyell and Hooker.

There is no evidence that Wallace thought his paper should have been handled differently.[8] He was told of the arrangement only after the fact, and yet he could not have been more gracious in thanking Darwin, Hooker, and Lyell for the steps they had taken in publishing his paper. In retrospect, Wallace wrote that "it was really a singular piece of good luck that gave to me any share whatever in the discovery" of natural selection because Darwin could have published his views at any point in the preceding twenty years.[9] Cognizant of the import of the theory, Wallace later expressed only praise for the book that he had spurred Darwin into publishing, and he was perfectly willing to differ with him on particular applications of the theory.

What Darwin understood immediately on reading Wallace's essay may not be as obvious to present-day readers. Wallace attempted, as was his habit, to lay out an argument as a logical sequence, but in this paper we find a tangle of subarguments about the differences between domesticated and wild animals, the causes of the relative abundance and scarcity of species, and the analogy between an individual's struggle for existence and the struggle among varieties of the same species. Wallace believed he had found the law that explained the origin of species. He was flush with malarial fever as well as intellectual fervor and wrote the essay with haste.

The theme of the essay is that animal species in nature are not permanent, which was a direct attack on the pervasive belief in the "fixity" of species, the idea that species were created once and for all and did not change over time. Wallace's argument takes some interesting twists and turns, but for the sake of clarity, it can be summarized as follows. Wallace had observed that variations exist among individuals of a species, and others, notably Malthus and Lyell, had pointed to a struggle for existence among individuals. When he combined these facts with the observations that more individuals are born than survive, he was led to the question of which ones survive. And the answer followed: those with variations better adapted to the vicissitudes of life, including competition, predation, and changing physical conditions. Following this insight, Wallace realized that there is a tendency for new varieties to continue to adapt to changing environments and to differ (diverge) from their ancestor. Here was Darwin's thesis in a nutshell; no wonder he was shocked and dismayed.

Wallace argued that this process (which came to be known as *evolution by natural selection*) applies to individuals in a species, to groups of allied species, and to varieties of a single species. In current language, he moved the unit of selection from individuals to species and then to varieties within a species, in each case making the same argument about variation, struggle, adaptation, and divergence. He also showed that domestic animals are not subject to a true struggle for survival, and therefore variations tend to revert to the original type rather than to diverge from it. The somewhat subdued language of his concluding paragraph belies the vision he had, for he had glimpsed a mechanism that explains the succession of all life, and he knew it.

❧ ❧
Collecting Birds of Paradise

My boat was at length ready, and having obtained two men besides my own servants, after an enormous amount of talk and trouble, we left Dobbo on the morning of March 13th, for the mainland of Aru [Figure 15]. By noon we reached the mouth of a small river or creek, which we ascended, winding among mangrove swamps, with here and there a glimpse of dry land. In two hours we reached a house, or rather small shed, of the most miserable description, which our steersman, the "Orang-kaya" of Wamma, said was the place we were to stay at, and where he had assured me we could get every kind of bird and beast to be found in Aru. The shed was occupied by about a dozen men, women, and children; two cooking fires were burning in it, and there seemed little prospect of my obtaining any accommodation. I however deferred inquiry till I had seen the neighbouring forest, and immediately started off with two men, net, and guns, along a path at the back of the house. In an hour's walk I saw enough to make me determine to give the place a trial, and on my return, finding the "Orang-kaya" was in a strong fever-fit and unable to do anything, I entered into negotiations with the owner of the house for the use of a slip at one end of it about five feet wide, for a week, and agreed to pay as rent one "parang," or chopping-knife. I then immediately got my boxes and bedding out of the boat, hung up a shelf for my bird-skins and insects, and got all ready for work next morning. My own boys slept in the boat to guard the remainder of my property; a cooking place sheltered by a few mats was arranged under a tree close by, and I felt that degree of satisfaction and enjoyment which I always experience when, after much trouble and delay, I am on the point of beginning work in a new locality.

One of my first objects was to inquire for the people who are accustomed to shoot the Paradise birds. They lived at some distance in the jungle, and a man was sent to call them. When they arrived, we had a talk by means of the "Orang-kaya" as interpreter, and they said they thought they could get some. They explained that they shoot the birds with a bow and arrow, the arrow having a conical wooden cap fitted to the end as large as a teacup,

The Malay Archipelago, chap. 31.

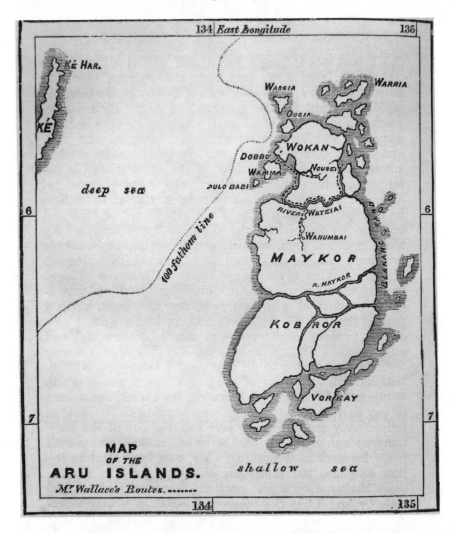

Fig. 15. Map of the Aru Islands. From *The Malay Archipelago*, p. 446.

so as to kill the bird by the violence of the blow without making any wound or shedding any blood. The trees frequented by the birds are very lofty; it is therefore necessary to erect a small leafy covering or hut among the branches, to which the hunter mounts before daylight in the morning and remains the whole day, and whenever a bird alights they are almost sure of securing it [see Figure 20]. They returned to their homes the same evening, and I never saw anything more of them, owing, as I afterwards found, to its being too early to obtain birds in good plumage.

Fig. 16. The king and the twelve-wired birds of paradise. From *The Malay Archipelago*, facing p. 552; original wood engraving by J. G. Keulemans.

The first two or three days of our stay here were very wet, and I obtained but few insects or birds, but at length, when I was beginning to despair, my boy Baderoon returned one day with a specimen which repaid me for months of delay and expectation. It was a small bird, a little less than a thrush. The greater part of its plumage was of an intense cinnabar red, with a gloss as of spun glass. On the head the feathers became short and velvety, and shaded into rich orange. Beneath, from the breast downwards, was pure white, with the softness and gloss of silk, and across the breast a band of deep metallic green separated this colour from the red of the throat. Above each eye was a round spot of the same metallic green; the bill was yellow, and the feet and legs were of a fine cobalt blue, strikingly contrasting with all the other parts of the body. Merely in arrangement of colours and texture of plumage this little bird was a gem of the first water, yet these comprised only half its strange beauty. Springing from each side of the breast, and ordinarily lying concealed under the wings, were little tufts of greyish feathers about two inches long, and each terminated by a broad band of intense emerald green. These plumes can be raised at the will of the bird, and spread out into a pair of elegant fans when the wings are elevated. But this is not the only ornament. The two middle feathers of the tail are in the form of slender wires about five inches long, and which diverge in a beautiful double curve. About half an inch of the end of this wire is webbed on the outer side only, and coloured of a fine metallic green, and being curled spirally inwards form a pair of elegant glittering buttons, hanging five inches below the body, and the same distance apart. These two ornaments, the breast fans and the spiral tipped tail wires, are altogether unique, not occurring on any other species of the eight thousand different birds that are known to exist upon the earth; and, combined with the most exquisite beauty of plumage, render this one of the most perfectly lovely of the many lovely productions of nature. My transports of admiration and delight quite amused my Aru hosts, who saw nothing more in the "Burong raja" than we do in the robin or the goldfinch. [The upper drawing in Figure 16 is a king bird of paradise, or *Paradisea regia* of Linneaus. Upon further study, its distinctive characteristics warranted making a new genus, and its name was changed to *Cicinnurus regius*.]

Thus one of my objects in coming to the far East was accomplished. I had obtained a specimen of the King Bird of Paradise (*Paradisea regia*), which had been described by Linnæus from skins preserved in a mutilated state by the natives. I knew how few Europeans had ever beheld the perfect little organism I now gazed upon, and how very imperfectly it was still known in Europe. The emotions excited in the minds of a naturalist, who has long desired to see the actual thing which he has hitherto known only by description, drawing, or badly-preserved external covering—especially when that thing is of surpassing rarity and beauty, require the poetic faculty fully to express them. The remote island in which I found myself situated, in an almost unvisited sea, far from the tracks of merchant fleets and navies; the wild luxuriant tropical forest, which stretched far away on every side; the rude uncultured savages who gathered round me,—all had their influence in determining the emotions with which I gazed upon this "thing of beauty." I thought of the long ages of the past, during which the successive generations of this little creature had run their course—year by year being born, and living and dying amid these dark and gloomy woods, with no intelligent eye to gaze upon their loveliness; to all appearance such a wanton waste of beauty. Such ideas excite a feeling of melancholy. It seems sad, that on the one hand such exquisite creatures should live out their lives and exhibit their charms only in these wild inhospitable regions, doomed for ages yet to come to hopeless barbarism; while on the other hand, should civilized man ever reach these distant lands, and bring moral, intellectual, and physical light into the recesses of these virgin forests, we may be sure that he will so disturb the nicely-balanced relations of organic and inorganic nature as to cause the disappearance, and finally the extinction, of these very beings whose wonderful structure and beauty he alone is fitted to appreciate and enjoy. This consideration must surely tell us that all living things were *not* made for man. Many of them have no relation to him. The cycle of their existence has gone on independently of his, and is disturbed or broken by every advance in man's intellectual development; and their happiness and enjoyments, their loves and hates, their struggles for existence, their vigorous life and early death, would seem to be immediately related to their own well-being and perpetuation alone, limited only by the equal well-being and perpetuation of the numberless other organisms with which each is more or less intimately connected.

After the first king-bird was obtained, I went with my men into the forest, and we were not only rewarded with another in equally perfect plumage, but I was enabled to see a little of the habits of both it and the larger species. It frequents the lower trees of the less dense forests, and is very active, flying strongly with a whirring sound, and continually hopping or flying from branch to branch. It eats hard stone-bearing fruits as large as a gooseberry, and often flutters its wings after the manner of the South American manakins, at which time it elevates and expands the beautiful fans with which its breast is adorned. The natives of Aru call it "Goby-goby."

One day I got under a tree where a number of the Great Paradise birds were assembled, but they were high up in the thickest of the foliage, and flying and jumping about so continually that I could get no good view of them. At length I shot one, but it was a young specimen, and was entirely of a rich chocolate-brown colour, without either the metallic green throat or yellow plumes of the full-grown bird. All that I had yet seen resembled this, and the natives told me that it would be about two months before any would be found in full plumage. I still hoped, therefore, to get some. Their voice is most extraordinary. At early morn, before the sun has risen, we hear a loud cry of "Wawk—wawk—wawk, wŏk—wŏk—wŏk," which resounds through the forest, changing its direction continually. This is the Great Bird of Paradise going to seek his breakfast. Others soon follow his example; lories and parroquets cry shrilly, cockatoos scream, king-hunters croak and bark, and the various smaller birds chirp and whistle their morning song. As I lie listening to these interesting sounds, I realize my position as the first European who has ever lived for months together in the Aru islands, a place which I had hoped rather than expected ever to visit. I think how many besides myself have longed to reach these almost fairy realms, and to see with their own eyes the many wonderful and beautiful things which I am daily encountering. But now Ali and Baderoon are up and getting ready their guns and ammunition, and little Baso has his fire lighted and is boiling my coffee, and I remember that I had a black cockatoo brought in late last night, which I must skin immediately, and so I jump up and begin my day's work very happily.

This cockatoo is the first I have seen, and is a great prize. It has a rather small and weak body, long weak legs, large wings, and an enormously developed head, ornamented with a magnificent crest, and armed with a sharp-pointed hooked bill of immense size and strength [Figure 17]. The plumage is entirely black, but has all over it the curious powdery white secretion characteristic of cockatoos. The cheeks are bare, and of an intense blood-red colour. Instead of the harsh scream of the white cockatoos, its voice is a somewhat plaintive whistle. The tongue is a curious organ, being

Fig. 17. Head of black cockatoo. From *The Malay Archipelago,*
p. 451; original wood engraving by T. J. Wood.

a slender fleshy cylinder of a deep red colour, terminated by a horny black plate, furrowed across and somewhat prehensile. The whole tongue has a considerable extensile power. I will here relate something of the habits of this bird, with which I have since become acquainted. It frequents the lower parts of the forest, and is seen singly, or at most two or three together. It flies slowly and noiselessly, and may be killed by a comparatively slight wound. It eats various fruits and seeds, but seems more particularly attached to the kernel of the kanary-nut, which grows on a lofty forest tree (*Canarium commune*), abundant in the islands where this bird is found;

and the manner in which it gets at these seeds shows a correlation of structure and habits, which would point out the "kanary" as its special food. The shell of this nut is so excessively hard that only a heavy hammer will crack it; it is somewhat triangular, and the outside is quite smooth. The manner in which the bird opens these nuts is very curious. Taking one endways in its bill and keeping it firm by a pressure of the tongue, it cuts a transverse notch by a lateral sawing motion of the sharp-edged lower mandible. This done, it takes hold of the nut with its foot, and biting off a piece of leaf retains it in the deep notch of the upper mandible, and again seizing the nut, which is prevented from slipping by the elastic tissue of the leaf, fixes the edge of the lower mandible in the notch, and by a powerful nip breaks off a piece of the shell. Again taking the nut in its claws, it inserts the very long and sharp point of the bill and picks out the kernel, which is seized hold of, morsel by morsel, by the extensible tongue. Thus every detail of form and structure in the extraordinary bill of this bird seems to have its use, and we may easily conceive that the black cockatoos have maintained themselves in competition with their more active and more numerous white allies, by their power of existing on a kind of food which no other bird is able to extract from its stony shell. The species is the *Microglossum aterrimum* of naturalists.

During the two weeks which I spent in this little settlement, I had good opportunities of observing the natives at their own home, and living in their usual manner. There is a great monotony and uniformity in every-day savage life, and it seemed to me a more miserable existence than when it had the charm of novelty. To begin with the most important fact in the existence of uncivilized peoples—their food—the Aru men have no regular supply, no staff of life, such as bread, rice, mandiocca, maize, or sago, which are the daily food of a large proportion of mankind. They have, however, many sorts of vegetables, plantains, yams, sweet potatoes, and raw sago; and they chew up vast quantities of sugar-cane, as well as betelnuts, gambir, and tobacco. Those who live on the coast have plenty of fish; but when inland, as we are here, they only go to the sea occasionally, and then bring home cockles and other shell-fish by the boatload. Now and then they get wild pig or kangaroo, but too rarely to form anything like a regular part of their diet, which is essentially vegetable; and what is of more importance, as affecting their health, green, watery vegetables, imperfectly cooked, and even these in varying and often insufficient quantities. To this diet may be attributed the prevalence of skin diseases, and ulcers on the legs and joints. The scurfy skin disease so common among savages has a close connexion with the poorness and irregularity of their living. The Malays, who are never without their daily rice, are generally free from it; the hill-Dyaks of Borneo, who grow rice and live well, are clean skinned, while the less industrious and less cleanly tribes, who live for a portion of the year on fruits and vegetables only, are very subject to this malady. It seems clear that in this, as in other respects, man is not able to make a beast of himself with impunity, feeding like the cattle on the herbs and fruits of the earth, and taking no thought of the morrow. To maintain his health and beauty he must labour to prepare some farinaceous product capable of being stored and accumulated, so as to give him a regular supply of wholesome food. When this is obtained, he may add vegetables, fruits, and meat with advantage.

The chief luxury of the Aru people, besides betel and tobacco, is arrack (Java rum), which the traders bring in great quantities and sell very cheap. A day's fishing or rattan cutting will purchase at least a half-gallon bottle; and when the tripang or birds' nests collected during a season are sold, they get whole boxes, each containing fifteen such bottles, which the inmates of a house will sit round day and night till they have finished. They themselves tell me that at such bouts they often tear to pieces the house they are in, break and destroy everything they can lay their hands on, and make such an infernal riot as is alarming to behold.

The houses and furniture are on a par with the food. A rude shed, supported on rough and slender sticks rather than posts, no walls, but the floor raised to within a foot of the eaves, is the style of architecture they usually adopt. Inside there are partition walls of thatch, forming little boxes or sleeping places, to accommodate the two or three separate families that usually live together. A few mats, baskets, and cooking vessels, with plates and basins purchased from the Macassar traders, constitute their whole furniture; spears and bows are their weapons; a sarong or mat forms the clothing of the women, a waistcloth of the men. For hours or even for days they sit idle in their houses, the women bringing in the vegetables or sago which form their food. Sometimes they hunt or fish a little, or work at their houses or canoes, but they seem to enjoy pure idleness, and work as little as they can. They have little to vary the monotony of life, little that can be called pleasure, except idleness and conversation. And they certainly do talk! Every evening there is a little Babel around me: but as I understand not a word of it, I go on with my book or work undisturbed. Now and then they scream and shout, or laugh frantically for variety; and this goes on alternately with vociferous talking of men, women, and children, till long after I am in my mosquito curtain and sound asleep [see Figure 18].

At this place I obtained some light on the complicated mixture of races in Aru, which would utterly confound an ethnologist. Many of the natives, though equally dark with the others, have little of the Papuan physiognomy, but have more delicate features of the European type, with more glossy, curling hair. These at first quite puzzled me, for they have no more resemblance to Malay than to Papuan, and the darkness of skin and hair would forbid the idea of Dutch intermixture. Listening to their conversation, however, I detected some words that were familiar to me. "Accabó" was one; and to be sure that it was not an accidental resemblance, I asked the speaker in Malay what *accabó*" meant, and was told it meant "done or finished," a true Portuguese word, with its meaning retained. Again, I heard the word "jafui" often repeated, and could see, without inquiry, that its meaning was "he's gone," as in Portuguese. "Porco," too, seems a common name, though the people have no idea of its European meaning. This cleared up the difficulty. I at once understood that some early Portuguese traders had penetrated to these islands, and mixed with the natives, influencing their language, and leaving in their descendants for many generations the visible characteristics of their race. If to this we add the occasional mixture of Malay, Dutch, and Chinese with the indigenous Papuans, we have no reason to wonder at the curious varieties of form and feature occasionally to be met with in Aru. In this very house there was a Macassar man, with an Aru wife and a family of mixed children. In Dobbo I saw a Javanese and an Amboyna man, each with an Aru wife and family; and as this kind of mixture has been going on for at least three hundred years, and probably much longer, it has produced a decided effect on the physical characteristics of a considerable portion of the population of the islands, more especially in Dobbo and the parts nearest it.

March 28th—The "Orang-kaya" [steersman] being very ill with fever had begged to go home, and had arranged with one of the men of the house to go on with me as his substitute. Now that I wanted to move, the bugbear of the pirates was brought up, and it was pronounced unsafe to go further than the next small river. This would not suit me, as I had determined to traverse the channel called Watelai to the "blakang-tana"; but my guide was firm in his dread of pirates, of which I knew there was now no danger, as several vessels had gone in search of them, as well as a Dutch gunboat which had arrived since I left Dobbo. I had, fortunately, by this time heard that the Dutch "Commissie" had really arrived, and therefore threatened that if my guide did not go with me immediately, I would appeal to the authorities, and he would certainly be obliged to give back the cloth which the "Orang-kaya" had transferred to him in prepayment. This had the desired effect; matters were soon arranged, and we started the next morning. The wind, however, was dead against us, and after rowing hard till midday we put in to a small river where there were a few huts, to cook our dinners. The place did not look very promising, but as we could not reach our destination, the Watelai river, owing to the contrary wind, I thought we might as well wait here a day or two. I therefore paid a chopper for the use of a small shed, and got my bed and some boxes on shore. In the evening, after dark, we were suddenly alarmed by the cry of "Bajak! bajak!" (Pirates!) The men all seized their bows and spears, and rushed down to the beach; we got hold of our guns and prepared for action, but in a few minutes all came back laughing and chattering, for it had proved to be only a small boat and some of their own comrades returned from fishing. When all was quiet again, one of the men, who could speak a little Malay, came to me and begged me not to sleep too hard. "Why?" said I. "Perhaps the pirates may really come," said he very seriously, which made me laugh and assure him I should sleep as hard as I could.

Two days were spent here, but the place was unproductive of insects or birds of interest, so we made another attempt to get on. As soon as we got a little away from the land we had a fair wind, and in six hours' sailing reached the entrance of the Watelai channel, which divides the most northerly from the middle portion of Aru. At its mouth this was about half a mile wide, but soon narrowed, and a mile or two on it assumed entirely the aspect of a river about the width of the Thames at London, winding among low but undulating and often hilly country. The scene was exactly such as might be expected in the interior of a continent. The channel continued of a uniform average width, with reaches and sinuous bends, one bank being often precipitous, or even forming vertical cliffs, while the other was flat and apparently alluvial; and it was only the pure salt-water, and the absence of any stream but the slight flux and reflux of the tide, that would enable a person to tell that he was navigating a strait and not a river. The wind was fair, and carried us along, with occasional assistance from our oars, till about three in the afternoon, when we landed where a little brook formed two or three basins in the coral rock, and then fell in a miniature cascade into the salt-water river. Here we bathed and cooked our dinner, and enjoyed ourselves lazily till sunset, when we pursued our way for two hours more, and then moored our little vessel to an overhanging tree for the night.

At five the next morning we started again, and in an hour overtook four large praus containing the "Commissie," who had come from Dobbo to make their official tour round the islands, and had passed us in the night. I paid a visit to the Dutchmen, one of whom spoke a little English, but we found that we could get on much better with Malay. They told me that they had been delayed going after the pirates to one of the northern islands, and had seen three of their vessels but could not catch them, because on being pursued they rowed out in the wind's eye, which they are enabled to do by having about fifty oars to each boat. Having had some tea with them, I bade them adieu, and turned up a narrow channel which our pilot said would take us to the village of Watelai, on the east side of Aru. After going some miles we found the channel nearly blocked up with coral, so that our boat grated along the bottom, crunching what may truly be called the living rock. Sometimes all hands had to get out and wade, to lighten the vessel and lift it over the shallowest places: but at length we overcame all obstacles and reached a wide bay or estuary studded with little rocks and islets, and opening to the eastern sea and the numerous islands of the "blakang-tana." I now found that the village we were going to was miles away; that we should have to go out to sea, and round a rocky point. A squall seemed coming on, and as I have a horror of small boats at sea, and from all I could learn Watelai village was not a place to stop at (no Birds of Paradise being found there), I determined to return and go to a village I had heard of up a tributary of the Watelai river, and situated nearly in the centre of the mainland of Aru. The people there were said to be good, and to be accustomed to hunting and bird-catching, being too far inland to get any part of their food from the sea. While I was deciding this point the squall burst upon us, and soon raised a rolling sea in the shallow water, which upset an oil bottle and a lamp, broke some of my crockery, and threw us all into confusion. Rowing hard we managed to get back into the main river by dusk, and looked out for a place to cook our suppers. It happened to be high water, and a very high tide, so that every piece of sand or beach was covered, and it was with the greatest difficulty, and after much groping in the dark, that we discovered a little sloping piece of rock about two feet square on which to make a fire and cook some rice. The next day we continued our way back, and on the following day entered a stream on the south side of the Watelai river, and ascending to where navigation ceased found the little village of Wanumbai, consisting of two large houses surrounded by plantations, amid the virgin forests of Aru.

As I liked the look of the place, and was desirous of staying some time, I sent my pilot to try and make a bargain for house accommodation. The owner and chief man of the place made many excuses. First, he was afraid I would not like his house, and then was doubtful whether his son, who was away, would like his admitting me. I had a long talk with him myself, and tried to explain what I was doing, and how many things I would buy of them, and showed him my stock of beads, and knives, and cloth, and tobacco, all of which I would spend with his family and friends if he would give me houseroom. He seemed a little staggered at this, and said he would talk to his wife, and in the meantime I went for a little walk to see the neighbourhood. When I came back, I again sent my pilot, saying that I would go away if he would not give me part of his house. In about half an hour he returned with a demand for about half the cost of building a house, for the rent of a small portion of it for a few weeks. As the only difficulty now was a pecuniary one, I got out about ten yards of cloth, an axe, with a few beads and some tobacco, and sent them as my final offer for the part of the house which I had before pointed out. This was accepted after a little more talk, and I immediately proceeded to take possession.

The house was a good large one, raised as usual about seven feet on posts, the walls about three or four feet more, with a high-pitched roof. The floor was of bamboo laths, and in the sloping roof was an immense shutter, which could be lifted and propped up to admit light and air. At the end where this was situated the floor was raised about a foot, and this piece, about ten feet wide by twenty long, quite open to the rest of the house, was the portion I was to occupy. At one end of this piece, separated by a thatch partition, was a cooking place, with a clay floor and shelves for crockery. At the opposite end I had my mosquito curtain hung, and round the walls we arranged my boxes and other stores, fitted up a table and seat, and with

Fig. 18. Native house from Aru. "From a photograph of a native house in the island of Wokan, which was given me by the late Professor Moseley of the *Challenger* expedition, because it so closely resembles the hut in which I lived for a fortnight, and where I obtained my first King bird of paradise, that I feel sure it must be the same, especially as I saw no other like it."
From *My Life*, 1:357; facing p. 356.

a little cleaning and dusting made the place look quite comfortable. My boat was then hauled up on shore, and covered with palm-leaves, the sails and oars brought indoors, a hanging-stage for drying my specimens erected outside the house and another inside, and my boys were set to clean their guns and get all ready for beginning work.

The next day I occupied myself in exploring the paths in the immediate neighbourhood. The small river up which we had ascended ceases to be navigable at this point, above which it is a little rocky brook, which quite dries up in the hot season. There was now, however, a fair stream of water in it; and a path which was partly in and partly by the side of the water, promised well for insects, as I here saw the magnificent blue butterfly, *Papilio ulysses*, as well as several other fine species, flopping lazily along, sometimes resting high up on the foliage which drooped over the water, at others settling down on the damp rock or on the edges of muddy pools. A little way on several paths branched off through patches of second-growth forest to cane-fields, gardens, and scattered houses, beyond which again the dark wall of verdure striped with tree-trunks, marked out the limits of the primeval forests. The voices of many birds promised good shooting, and on my return I found that my boys had already obtained two or three kinds I had not seen before; and in the evening a native brought me a rare and beautiful species of ground-thrush (*Pitta novæ-guineæ*) hitherto only known from New Guinea.

As I improved my acquaintance with them I became much interested in these people, who are a fair sample of the true savage inhabitants of the Aru Islands, tolerably free from foreign admixture. The house I lived in contained four or five families, and there were generally from six to a dozen visitors besides. They kept up a continual row from morning till night—talking, laughing, shouting, without intermission—not very pleasant, but interesting as a study of national character. My boy Ali said to me, "Banyak quot bitchara Orang Aru" (The Aru people are very strong talkers), never having been accustomed to such eloquence either in his own or any other country he had hitherto visited. Of an evening the men, having got over their first shyness, began to talk to me a little, asking about my country, &c., and in return I questioned them about any traditions they had of their own origin. I had, however, very little success, for I could not possibly make them understand the simple question of where the Aru people first came from. I put it in every possible way to them, but it was a subject quite beyond their speculations; they had evidently never thought of anything of the kind, and were unable to conceive a thing so remote and so unnecessary to be thought about, as their own origin. Finding this hopeless, I asked if they knew when the trade with Aru first began, when the Bugis and Chinese and Macassar men first came in their praus to buy tripang and tortoise-shell, and birds' nests, and Paradise birds? This they comprehended, but replied that there had always been the same trade as long as they or their fathers recollected, but that this was the first time a real white man had come among them, and, said they, "You see how the people come every day from all the villages round to look at you." This was very flattering, and accounted for the great concourse of visitors which I had at first imagined was accidental. A few years before I had been one of the gazers at the Zoolus and the Aztecs in London. Now the tables were turned upon me, for I was to these people a new and strange variety of man, and had the honour of affording to them, in my own person, an attractive exhibition, gratis.

All the men and boys of Aru are expert archers, never stirring without their bows and arrows. They shoot all sorts of birds, as well as pigs and kangaroos occasionally, and thus have a tolerably good supply of meat to eat with their vegetables. The result of this better living is superior healthiness, well-made bodies, and generally clear skins. They brought me numbers of small birds in exchange for beads or tobacco, but mauled them terribly, notwithstanding my repeated instructions. When they got a bird alive they would often tie a string to its leg, and keep it a day or two, till its plumage was so draggled and dirtied as to be almost worthless. One of the first things I got from them was a living specimen of the curious and beautiful racquet-tailed kingfisher. Seeing how much I admired it, they afterwards brought me several more, which were all caught before daybreak, sleeping in cavities of the rocky banks of the stream. My hunters also shot a few specimens, and almost all of them had the red bill more or less clogged with mud and earth. This indicates the habits of the bird, which, though popularly a kingfisher, never catches fish, but lives on insects and minute shells, which it picks up in the forest, darting down upon them from its perch on some low branch. The genus *Tanysiptera*, to which this bird belongs, is remarkable for the enormously lengthened tail, which in all other kingfishers is small and short. Linnæus named the species known to him "the goddess kingfisher" (*Alcedo dea*), from its extreme grace and beauty, the plumage being brilliant blue and white, with the bill red, like coral. Several species of these interesting birds are now known, all confined within the very limited area which comprises the Moluccas, New Guinea, and the extreme North of Australia. They resemble each other so closely that several of them can only be distinguished by careful comparison. One of the rarest, however, which inhabits New Guinea, is very distinct from the rest, being bright red beneath instead of white. That which I now obtained was a new one, and has been named *Tanysiptera hydrocharis*, but in general form and coloration it is exactly similar to the larger species found in Amboyna [see Figure 19].

New and interesting birds were continually brought in, either by my own boys or by the natives, and at the end of a week Ali arrived triumphant one afternoon with a fine specimen of the Great Bird of Paradise. The ornamental plumes had not yet attained their full growth, but the richness of their glossy orange colouring, and the exquisite delicacy of the loosely waving feathers, were unsurpassable. At the same time a great black cockatoo was brought in, as well as a fine fruit-pigeon and several small birds, so that we were all kept hard at work skinning till sunset. Just as we had cleared

Fig. 19. The racquet-tailed kingfisher. From *The Malay Archipelago,* facing p. 298; original wood engraving by T. J. Wood.

away and packed up for the night, a strange beast was brought, which had been shot by the natives. It resembled in size, and in its white woolly covering, a small fat lamb, but had short legs, hand-like feet with large claws, and a long prehensile tail. It was a *Cuscus* (*C. maculatus*), one of the curious marsupial animals of the Papuan region, and I was very desirous to obtain the skin. The owners, however, said they wanted to eat it; and though I offered them a good price, and promised to give them all the meat, there was great hesitation. Suspecting the reason, I offered, though it was night, to set to work immediately and get out the body for them, to which they agreed. The creature was much hacked about, and the two hind feet almost cut off, but it was the largest and finest specimen of the kind I had seen; and after an hour's hard work I handed over the body to the owners, who immediately cut it up and roasted it for supper.

As this was a very good place for birds, I determined to remain a month longer, and took the opportunity of a native boat going to Dobbo, to send Ali for a fresh supply of ammunition and provisions. They started on the 10th of April, and the house was crowded with about a hundred men, boys, women, and girls, bringing their loads of sugar-cane, plantains, sirih-leaf, yams, &c.; one lad going from each house to sell the produce and make purchases. The noise was indescribable. At least fifty of the hundred were always talking at once, and that not in the low measured tones of the apathetically polite Malay, but with loud voices, shouts, and screaming laughter, in which the women and children were even more conspicuous than the men. It was only while gazing at me that their tongues were moderately quiet, because their eyes were fully occupied. The black vegetable soil here overlying the coral rock is very rich, and the sugar-cane was finer than any I had ever seen. The canes brought to the boat were often ten and even twelve feet long, and thick in proportion, with short joints throughout, swelling between the knots with the abundance of the rich juice. At Dobbo they get a high price for it, 1*d.* to 3*d.* a stick, and there is an insatiable demand among the crews of the praus and the Baba fishermen. Here they eat it continually. They half live on it, and sometimes feed their pigs with it. Near every house are great heaps of the refuse cane; and large wicker-baskets to contain this refuse as it is produced form a regular part of the furniture of a house. Whatever time of the day you enter, you are sure to find three or four people with a yard of cane in one hand, a knife in the other, and a basket between their legs, hacking, paring, chewing, and basket-fill-ing, with a persevering assiduity which reminds one of a hungry cow grazing, or of a caterpillar eating up a leaf.

After five days' absence the boats returned from Dobbo, bringing Ali and all the things I had sent for quite safe. A large party had assembled to be ready to carry home the goods brought, among which were a good many cocoa-nuts, which are a great luxury here. It seems strange that they should never plant them; but the reason simply is, that they cannot bring their hearts to bury a good nut for the prospective advantage of a crop twelve years hence. There is also the chance of the fruits being dug up and eaten unless watched night and day. Among the things I had sent for was a box of arrack, and I was now of course besieged with requests for a little drop. I gave them a flask (about two bottles), which was very soon finished, and I was assured that there were many present who had not had a taste. As I feared my box would very soon be emptied if I supplied all their demands, I told them I had given them one, but the second they must pay for, and that afterwards I must have a Paradise bird for each flask. They immediately sent round to all the neighbouring houses, and mustered up a rupee in Dutch copper money, got their second flask, and drunk it as quickly as the first, and were then very talkative, but less noisy and importunate than I expected. Two or three of them got round me and begged me for the twentieth time to tell them the name of my country. Then, as they could not pronounce it satisfactorily, they insisted that I was deceiving them, and that it was a name of my own invention. One funny old man, who bore a ludicrous resemblance to a friend of mine at home, was almost indignant. "Unglung!" said he, "who ever heard of such a name?—anglang—anger-lang—that can't be the name of your country; you are playing with us." Then he tried to give a convincing illustration. "My country is Wanumbai—anybody can say Wanumbai. I'm an 'orang-Wanumbai'; but, N-glung! who ever heard of such a name? Do tell us the real name of your country, and then when you are gone we shall know how to talk about you." To this luminous argument and remonstrance I could oppose nothing but assertion, and the whole party remained firmly convinced that I was for some reason or other deceiving them. They then attacked me on another point—what all the animals and birds and insects and shells were preserved so carefully for. They had often asked me this before, and I had tried to explain to them that they would be stuffed, and made to look as if alive, and people in my country would go to look at them. But this was not satisfying; in my country there must be many better things to look at, and they could not believe I would take so much trouble with their birds and beasts just for people to look at. They did not want to look at them; and we, who made calico and glass and knives, and all sorts of wonderful things, could not want things from Aru to look at. They had evidently been thinking about it, and had at length got what seemed a very satisfactory theory; for the same old man said to me, in a low mysterious voice, "What becomes of them when you go on to the sea?" "Why, they are all packed up in boxes," said I. "What did you think became of them?" "They all come to life again, don't they?" said he; and though I tried to joke it off, and said if they did we should have plenty to eat at sea, he stuck to his opinion, and kept repeating, with an air of deep conviction, "Yes, they all come to life again, that's what they do—they all come to life again."

After a little while, and a good deal of talking among themselves, he began again—"I know all about it—oh, yes! Before you came we had rain every day—very wet indeed; now, ever since you have been here, it is fine hot weather. Oh yes! I know all about it; you can't deceive me." And so I was set down as a conjurer, and was unable to repel the charge. But the conjurer was completely puzzled by the next question: "What," said the old man, "is the great ship, where the Bugis and Chinamen go to sell their things? It is always in the great sea—its name is Jong; tell us all about it." In vain I inquired what they knew about it; they knew nothing but that it was called "Jong," and was always in the sea, and was a very great ship, and concluded with, "Perhaps that is your country?" Finding that I could not or would not tell them anything about "Jong," there came more regrets that I would not tell them the real name of my country; and then a long string of compliments, to the effect that I was a much better sort of a person than the Bugis and Chinese, who sometimes came to trade with them, for I gave them things for nothing, and did not try to cheat them. How long would I stop? was the next earnest inquiry. Would I stay two or three months? They would get me plenty of birds and animals, and I might soon finish all the goods I had brought, and then, said the old spokesman, "Don't go away, but send for more things from Dobbo, and stay here a year or two." And then again the old story, "Do tell us the name of your country. We know the Bugis men, and the Macassar men, and the Java men, and the Chinamen; only you, we don't know from what country you come. Ung-lung! it can't be; I know that is not the name of your country." Seeing no end to this long talk,

I said I was tired, and wanted to go to sleep; so after begging—one a little bit of dry fish for his supper, and another a little salt to eat with his sago— they went off very quietly, and I went outside and took a stroll round the house by moonlight, thinking of the simple people and the strange productions of Aru, and then turned in under my mosquito curtain, to sleep with a sense of perfect security in the midst of these good-natured savages.

We now had seven or eight days of hot and dry weather, which reduced the little river to a succession of shallow pools connected by the smallest possible thread of trickling water. If there were a dry season like that of Macassar, the Aru Islands would be uninhabitable, as there is no part of them much above a hundred feet high; and the whole being a mass of porous coralline rock, allows the surface water rapidly to escape. The only dry season they have is for a month or two about September or October, and there is then an excessive scarcity of water, so that sometimes hundreds of birds and other animals die of drought. The natives then remove to houses near the sources of the small streams, where, in the shady depths of the forest, a small quantity of water still remains. Even then many of them have to go miles for their water, which they keep in large bamboos and use very sparingly. They assure me that they catch and kill game of all kinds, by watching at the water holes or setting snares around them. That would be the time for me to make my collections; but the want of water would be a terrible annoyance, and the impossibility of getting away before another whole year had passed made it out of the question.

Ever since leaving Dobbo I had suffered terribly from insects, who seemed here bent upon revenging my long-continued persecution of their race. At our first stopping-place sand-flies were very abundant at night, penetrating to every part of the body, and producing a more lasting irritation than mosquitoes. My feet and ankles especially suffered, and were completely covered with little red swollen specks, which tormented me horribly. On arriving here we were delighted to find the house free from sand-flies or mosquitoes, but in the plantations where my daily walks led me, the day-biting mosquitoes swarmed, and seemed especially to delight in attacking my poor feet. After a month's incessant punishment, those useful members rebelled against such treatment and broke into open insurrection, throwing out numerous inflamed ulcers, which were very painful, and stopped me from walking. So I found myself confined to the house, and with no immediate prospect of leaving it. Wounds or sores in the feet are especially difficult to heal in hot climates, and I therefore dreaded them more than any other illness. The confinement was very annoying, as the fine hot weather was excellent for insects, of which I had every promise of obtaining a fine collection; and it is only by daily and unremitting search that the smaller kinds, and the rarer and more interesting specimens, can be obtained. When I crawled down to the river-side to bathe, I often saw the blue-winged *Papilio ulysses,* or some other equally rare and beautiful insect; but there was nothing for it but patience, and to return quietly to my bird-skinning, or whatever other work I had indoors. The stings and bites and ceaseless irritation caused by these pests of the tropical forests, would be borne uncomplainingly; but to be kept prisoner by them in so rich and unexplored a country, where rare and beautiful creatures are to be met with in every forest ramble—a country reached by such a long and tedious voyage, and which might not in the present century be again visited for the same purpose—is a punishment too severe for a naturalist to pass over in silence.

I had, however, some consolation in the birds my boys brought home daily, more especially the *Paradiseas,* which they at length obtained in full plumage. It was quite a relief to my mind to get these, for I could hardly have torn myself away from Aru had I not obtained specimens. But what I valued almost as much as the birds themselves was the knowledge of their habits, which I was daily obtaining both from the accounts of my hunters, and from the conversation of the natives. The birds had now commenced what the people here call their "sácaleli," or dancing-parties, in certain trees in the forest, which are not fruit trees as I at first imagined, but which have an immense head of spreading branches and large but scattered leaves, giving a clear space for the birds to play and exhibit their plumes. On one of these trees a dozen or twenty full-plumaged male birds assemble together, raise up their wings, stretch out their necks, and elevate their exquisite plumes, keeping them in a continual vibration. Between whiles they fly across from branch to branch in great excitement, so that the whole tree is filled with waving plumes in every variety of attitude and motion [Figure 20]. The bird itself is nearly as large as a crow, and is of a rich coffee brown colour. The head and neck is of a pure straw yellow above, and rich metallic green beneath. The long plumy tufts of golden orange feathers spring from the sides beneath each wing, and when the bird is in repose are partly concealed by them. At the time of its excitement, however, the

Fig. 20. Natives of Aru shooting the great bird of paradise. From *The Malay Archipelago,* frontispiece; original wood engraving by T. J. Wood.

wings are raised vertically over the back, the head is bent down and stretched out, and the long plumes are raised up and expanded till they form two magnificent golden fans striped with deep red at the base, and fading off into the pale brown tint of the finely divided and softly waving points. The whole bird is then overshadowed by them, the crouching body, yellow head, and emerald green throat forming but the foundation and setting to the golden glory which waves above. When seen in this attitude, the Bird of Paradise really deserves its name, and must be ranked as one of the most beautiful and most wonderful of living things. I continued also to get specimens of the lovely little king-bird occasionally, as well as numbers of brilliant pigeons, sweet little parroquets, and many curious small birds, most nearly resembling those of Australia and New Guinea.

ﻊ   Here, as among most savage people I have dwelt among, I was delighted with the beauty of the human form—a beauty of which stay-at-home civilized people can scarcely have any conception. What are the finest Grecian statues to the living, moving, breathing men I saw daily around me? The unrestrained grace of the naked savage as he goes about his daily occupations, or lounges at his ease, must be seen to be understood; and a youth bending his bow is the perfection of manly beauty. The women, however, except in extreme youth, are by no means so pleasant to look at as the men. Their strongly-marked features are very unfeminine, and hard work, privations, and very early marriages soon destroy whatever of beauty or grace they may for a short time possess. Their toilet is very simple, but also, I am sorry to say, very coarse, and disgusting. It consists solely of a mat of plaited strips of palm leaves, worn tight round the body, and reaching from the hips to the knees. It seems not to be changed till worn out, is seldom washed, and is generally very dirty. This is the universal dress, except in a few cases where Malay "sarongs" have come into use. Their frizzly hair is tied in a bunch at the back of the head. They delight in combing, or rather forking it, using for that purpose a large wooden fork with four diverging prongs, which answers the purpose of separating and arranging the long tangled, frizzly mass of cranial vegetation much better than any comb could do. The only ornaments of the women are earrings and necklaces, which they arrange in various tasteful ways. The ends of a necklace are often attached to the earrings, and then looped on to the hairknot behind. This has really an elegant appearance, the beads hanging gracefully on each side of the head, and by establishing a connexion with the earrings give an appearance of utility to those barbarous ornaments. We recommend this style to the consideration of those of the fair sex who still bore holes in their ears and hang rings thereto. Another style of necklace among these Papuan belles is to wear two, each hanging on one side of the neck and under the opposite arm, so as to cross each other. This has a very pretty appearance, in part due to the contrast of the white beads or kangaroo teeth of which they are composed with the dark glossy skin. The earrings themselves are formed of a bar of copper or silver, twisted so that the ends cross. The men, as usual among savages, adorn themselves more than the women. They wear necklaces, earrings, and finger rings, and delight in a band of plaited grass tight round the arm just below the shoulder, to which they attach a bunch of hair or bright coloured feathers by way of ornament. The teeth of small animals, either alone, or alternately with black or white beads, form their necklaces, and sometimes bracelets also. For these latter, however, they prefer brass wire, or the black, horny, wing-spines of the cassowary, which they consider a charm. Anklets of brass or shell, and tight plaited garters below the knee, complete their ordinary decorations.

Some natives of Kobror from further south, and who are reckoned the worst and least civilized of the Aru tribes, came one day to visit us. They have a rather more than usually savage appearance, owing to the greater amount of ornaments they use—the most conspicuous being a large horseshoe-shaped comb which they wear over the forehead, the ends resting on the temples. The back of the comb is fastened into a piece of wood, which is plated with tin in front, and above is attached a plume of feathers from a cock's tail. In other respects they scarcely differed from the people I was living with. They brought me a couple of birds, some shells and insects, showing that the report of the white man and his doings had reached their country. There was probably hardly a man in Aru who had not by this time heard of me.

Besides the domestic utensils already mentioned, the moveable property of a native is very scanty. He has a good supply of spears and bows and arrows for hunting, a parang, or chopping-knife, and an axe—for the stone age has passed away here, owing to the commercial enterprise of the Bugis and other Malay races. Attached to a belt, or hung across his shoulder, he carries a little skin pouch and an ornamented bamboo, containing betel-nut, tobacco, and lime, and a small German wooden-handled knife is generally stuck between his waist-cloth of bark and his bare skin. Each man also possesses a "cadjan," or sleeping-mat, made of the broad leaves of a pandanus neatly sewn together in three layers. This mat is about four feet square, and when folded has one end sewn up, so that it forms a kind of sack open at one side. In the closed corner the head or feet can be placed, or by carrying it on the head in a shower it forms both coat and umbrella. It doubles up in a small compass for convenient carriage, and then forms a light and elastic cushion, so that on a journey it becomes clothing, house, bedding, and furniture, all in one.

The only ornaments in an Aru house are trophies of the chase—jaws of wild pigs, the heads and backbones of cassowaries, and plumes made from the feathers of the Bird of Paradise, cassowary, and domestic fowl. The spears, shields, knife-handles, and other utensils are more or less carved in fanciful designs, and the mats and leaf boxes are painted or plaited in neat patterns of red, black, and yellow colours. I must not forget these boxes, which are most ingeniously made of the pith of a palm leaf pegged together, lined inside with pandanus leaves, and outside with the same, or with plaited grass. All the joints and angles are covered with strips of split rattan sewn neatly on. The lid is covered with the brown leathery spathe of the *Areca* palm, which is impervious to water, and the whole box is neat, strong, and well finished. They are made from a few inches to two or three feet long, and being much esteemed by the Malays as clothes-boxes, are a regular article of export from Aru. The natives use the smaller ones for tobacco or betel-nut, but seldom have clothes enough to require the larger ones, which are only made for sale.

Among the domestic animals which may generally be seen in native houses, are gaudy parrots, green, red, and blue, a few domestic fowls, which have baskets hung for them to lay in under the eaves, and who sleep on the ridge, and several half-starved wolfish-looking dogs. Instead of rats and mice there are curious little marsupial animals about the same size, which run about at night and nibble anything eatable that may be left uncovered. Four or five different kinds of ants attack everything not isolated by water, and one kind even swims across that; great spiders lurk in baskets and boxes, or hide in the folds of my mosquito curtain; centipedes and mille-pedes are found everywhere. I have caught them under my pillow and on my head; while in every box, and under every board which has lain for some days undisturbed, little scorpions are sure to be found snugly ensconced, with their formidable tails quickly turned up ready for attack or defence. Such companions seem very alarming and dangerous, but all combined are not so bad as the irritation of mosquitoes, or of the insect pests often found at home. These latter are a constant and unceasing source of torment and disgust, whereas you may live a long time among scorpions, spiders, and centipedes, ugly and venomous though they are, and get no harm from them. After living twelve years in the tropics, I have never yet been bitten or stung by either.

The lean and hungry dogs before mentioned were my greatest enemies, and kept me constantly on the watch. If my boys left the bird they were skinning for an instant, it was sure to be carried off. Everything eatable had to be hung up to the roof, to be out of their reach. Ali had just finished skinning a fine King Bird of Paradise one day, when he dropped the skin. Before he could stoop to pick it up, one of this famished race had seized upon it, and he only succeeded in rescuing it from its fangs after it was torn to tatters. Two skins of the large Paradisea, which were quite dry and ready to pack away, were incautiously left on my table for the night, wrapped up in paper. The next morning they were gone, and only a few scattered feathers indicated their fate. My hanging shelf was out of their reach; but having stupidly left a box which served as a step, a full-plumaged Paradise bird was next morning missing; and a dog below the house was to be seen still mumbling over the fragments, with the fine golden plumes all trampled in the mud. Every night, as soon as I was in bed, I could hear them searching about for what they could devour, under my table, and all about my boxes and baskets, keeping me in a state of suspense till morning, lest something of value might incautiously have been left within their reach. They would drink the oil of my floating lamp and eat the wick, and upset or break my crockery if my lazy boys had neglected to wash away even the smell of anything eatable. Bad, however, as they are here, they were worse in a Dyak's house in Borneo where I was once staying, for there they gnawed off the tops of my waterproof boots, ate a large piece out of an old leather game-bag, besides devouring a portion of my mosquito curtain!

April 28th—Last evening we had a grand consultation, which had evidently been arranged and discussed beforehand. A number of the natives gathered round me, and said they wanted to talk. Two of the best Malay scholars helped each other, the rest putting in hints and ideas in their own language. They told me a long rambling story; but, partly owing to their imperfect knowledge of Malay, partly through my ignorance of local terms, and partly through the incoherence of their narrative, I could not make it out very clearly. It was, however, a tradition, and I was glad to find they had anything of the kind. A long time ago, they said, some strangers came to Aru, and came here to Wanumbai, and the chief of the Wanumbai people did not like them, and wanted them to go away, but they would not go, and so it came to fighting, and many Aru men were killed, and some, along with the chief, were taken prisoners, and carried away by the strangers. Some of the speakers, however, said that he was not carried away, but went away in his own boat to escape from the foreigners, and went to the sea and never came back again. But they all believe that the chief and the people that went with him still live in some foreign country; and if they could but find out where, they would send for them to come back again. Now having some vague idea that white men must know every country beyond the sea, they wanted to know if I had met their people in my country or in the sea. They thought they must be there, for they could not imagine where else they could be. They had sought for them everywhere, they said—on the land and in the sea, in the forest and on the mountains, in the air and in the sky, and could not find them; therefore, they must be in my country, and they begged me to tell them, for I must surely know, as I came from across the great sea. I tried to explain to them that their friends could not have reached my country in small boats; and that there were plenty of islands like Aru all about the sea, which they would be sure to find. Besides, as it was so long ago, the chief and all the people must be dead. But they quite laughed at this idea, and said they were sure they were alive, for they had proof of it. And then they told me that a good many years ago, when the speakers were boys, some Wokan men who were out fishing met these lost people in the sea, and spoke to them; and the chief gave the Wokan men a hundred fathoms of cloth to bring to the men of Wanumbai, to show that they were alive and would soon come back to them; but the Wokan men were thieves, and kept the cloth, and they only heard of it afterwards; and when they spoke about it, the Wokan men denied it, and pretended they had not received the cloth;—so they were quite sure their friends were at that time alive and somewhere in the sea. And again, not many years ago, a report came to them that some Bugis traders had brought some children of their lost people; so they went to Dobbo to see about it, and the owner of the house, who was now speaking to me, was one who went; but the Bugis man would not let them see the children, and threatened to kill them if they came into his house. He kept the children shut up in a large box, and when he went away he took them with him. And at the end of each of these stories, they begged me in an imploring tone to tell them if I knew where their chief and their people now were.

By dint of questioning, I got some account of the strangers who had taken away their people. They said they were wonderfully strong, and each one could kill a great many Aru men; and when they were wounded, however badly, they spit upon the place, and it immediately became well. And they made a great net of rattans, and entangled their prisoners in it, and sunk them in the water; and the next day, when they pulled the net up on shore, they made the drowned men come to life again, and carried them away.

Much more of the same kind was told me, but in so confused and rambling a manner that I could make nothing out of it, till I inquired how long ago it was that all this happened, when they told me that after their people were taken away the Bugis came in their praus to trade in Aru, and to buy tripang and birds' nests. It is not impossible that something similar to what they related to me really happened when the early Portuguese discoverers first came to Aru, and has formed the foundation for a continually increasing accumulation of legend and fable. I have no doubt that to the next generation, or even before, I myself shall be transformed into a magician or a demigod, a worker of miracles, and a being of supernatural knowledge. They already believe that all the animals I preserve will come to life again; and to their children it will be related that they actually did so. An unusual spell of fine weather setting in just at my arrival has made them believe I can control the seasons; and the simple circumstance of my always walking alone in the forest is a wonder and a mystery to them, as well as my asking them about birds and animals I have not yet seen, and showing an acquaintance with their forms, colours, and habits. These facts are brought against me when I disclaim knowledge of what they wish me to tell them. "You must know," say they; "you know everything: you make the fine weather for your men to shoot; and you know all about our birds and our animals as well as we do; and you go alone into the forest and are not afraid." Therefore every confession of ignorance on my part is thought to be a blind, a mere excuse to avoid telling them too much. My very writing materials and books are to them weird things; and were I to choose to mystify them by a few simple experiments with lens and magnet, miracles without end would in a few years cluster about me; and future travellers, penetrating to Wanumbai, would hardly believe that a poor English naturalist, who had resided a few months among them, could have been the original of the supernatural being to whom so many marvels were attributed.

For some days I had noticed a good deal of excitement, and many strangers came and went armed with spears and cutlasses, bows and shields. I now found there was war near us—two neighbouring villages having a quarrel about some matter of local politics that I could not understand. They told me it was quite a common thing, and that they are rarely without fighting somewhere near. Individual quarrels are taken up by villages and tribes, and the nonpayment of the stipulated price for a wife is one of the most frequent causes of bitterness and bloodshed. One of the war shields was brought me to look at. It was made of rattans and covered with cotton twist, so as to be both light, strong, and very tough. I should think it would resist an ordinary bullet. About the middle there was an arm-hole with a shutter or flap over it. This enables the arm to be put through and the bow drawn, while the body and face, up to the eyes, remain protected, which cannot be done if the shield is carried on the arm by loops attached at the back in the ordinary way. A few of the young men from our house went to help their friends, but I could not hear that any of them were hurt, or that there was much hard fighting.

May 8th—I had now been six weeks at Wanumbai, but for more than half the time was laid up in the house with ulcerated feet. My stores being nearly exhausted, and my bird and insect boxes full, and having no immediate prospect of getting the use of my legs again, I determined on returning to Dobbo. Birds had lately become rather scarce, and the Paradise birds had not yet become as plentiful as the natives assured me they would be in another month. The Wanumbai people seemed very sorry at my departure; and well they might be, for the shells and insects they picked up on the way to and from their plantations, and the birds the little boys shot with their bows and arrows, kept them all well supplied with tobacco and gambir, besides enabling them to accumulate a stock of beads and coppers for future expenses. The owner of the house was supplied gratis with a little rice, fish, or salt, whenever he asked for it, which I must say was not very often. On parting, I distributed among them my remnant stock of salt and tobacco, and gave my host a flask of arrack, and believe that on the whole my stay with these simple and good-natured people was productive of plea-sure and profit to both parties. I fully intended to come back; and had I known that circumstances would have prevented my doing so, should have felt some sorrow in leaving a place where I had first seen so many rare and beautiful living things, and had so fully enjoyed the pleasure which fills the heart of the naturalist when he is so fortunate as to discover a district hith-erto unexplored, and where every day brings forth new and unexpected treasures. We loaded our boat in the afternoon, and, starting before day-break, by the help of a fair wind reached Dobbo late the same evening.

On the Tendency of Varieties to Depart Indefinitely from the Original Type

One of the strongest arguments which have been adduced to prove the orig-inal and permanent distinctness of species is, that *varieties* produced in a state of domesticity are more or less unstable, and often have a tendency, if left to themselves, to return to the normal form of the parent species; and this instability is considered to be a distinctive peculiarity of all varieties, even of those occurring among wild animals in a state of nature, and to constitute a provision for preserving unchanged the originally created dis-tinct species.

In the absence or scarcity of facts and observations as to *varieties* oc-curring among wild animals, this argument has had great weight with nat-uralists, and has led to a very general and somewhat prejudiced belief in the stability of species. Equally general, however, is the belief in what are called "permanent or true varieties"—races of animals which continually propagate their like, but which differ so slightly (although constantly) from some other race, that the one is considered to be a *variety* of the other. Which is the *variety* and which the original *species*, there is generally no means of determining, except in those rare cases in which the one race has been known to produce an offspring unlike itself and resembling the other. This, however, would seem quite incompatible with the "permanent in-variability of species," but the difficulty is overcome by assuming that such

Proceedings of the Linnean Society of London 3, no. 9 (1858):53–62.

varieties have strict limits, and can never again vary further from the original type, although they may return to it, which, from the analogy of the domesticated animals, is considered to be highly probable, if not certainly proved.

It will be observed that this argument rests entirely on the assumption, that *varieties* occurring in a state of nature are in all respects analogous to or even identical with those of domestic animals, and are governed by the same laws as regards their permanence or further variation. But it is the object of the present paper to show that this assumption is altogether false, that there is a general principle in nature which will cause many *varieties* to survive the parent species, and to give rise to successive variations departing further and further from the original type, and which also produces, in domesticated animals, the tendency of varieties to return to the parent form.

The life of wild animals is a struggle for existence. The full exertion of all their faculties and all their energies is required to preserve their own existence and provide for that of their infant offspring. The possibility of procuring food during the least favourable seasons, and of escaping the attacks of their most dangerous enemies, are the primary conditions which determine the existence both of individuals and of entire species. These conditions will also determine the population of a species; and by a careful consideration of all the circumstances we may be enabled to comprehend, and in some degree to explain, what at first sight appears so inexplicable—the excessive abundance of some species, while others closely allied to them are very rare.

The general proportion that must obtain between certain groups of animals is readily seen. Large animals cannot be so abundant as small ones; the Carnivora must be less numerous than the Herbivora; eagles and lions can never be so plentiful as pigeons and antelopes; the wild asses of the Tartarian deserts cannot equal in numbers the horses of the more luxuriant prairies and pampas of America. The greater or less fecundity of an animal is often considered to be one of the chief causes of its abundance or scarcity; but a consideration of the facts will show us that it really has little or nothing to do with the matter. Even the least prolific of animals would increase rapidly if unchecked, whereas it is evident that the animal population of the globe must be stationary, or perhaps, through the influence of man, decreasing. Fluctuations there may be; but permanent increase, except in restricted localities, is almost impossible. For example, our own observation must convince us that birds do not go on increasing every year in a geometrical ratio, as they would do, were there not some powerful check to their natural increase. Very few birds produce less than two young ones each year, while many have six, eight or ten; four will certainly be below the average; and if we suppose that each pair produce young only four times in their life, that will also be below the average, supposing them not to die either by violence or want of food. Yet at this rate how tremendous would be the increase in a few years from a single pair! A simple calculation will show that in fifteen years each pair of birds would have increased to nearly ten millions! whereas we have no reason to believe that the number of the birds of any country increases at all in fifteen or in one hundred and fifty years. With such powers of increase the population must have reached its limits, and have become stationary, in a very few years after the origin of each species. It is evident, therefore, that each year an immense number of birds must perish—as many in fact as are born; and as on the lowest calculation the progeny are each year twice as numerous as their parents, it follows that, whatever be the average number of individuals existing in any given country, *twice that number must perish annually*—a striking result, but one which seems at least highly probable, and is perhaps under rather than over the truth. It would therefore appear that, as far as the continuance of the species and the keeping up the average number of individuals are concerned, large broods are superfluous. On the average all above *one* become food for hawks and kites; wild cats and weasels, or perish of cold and hunger as winter comes on. This is strikingly proved by the case of particular species; for we find that their abundance in individuals bears no relation whatever to their fertility in producing offspring. Perhaps the most remarkable instance of an immense bird population is that of the passenger pigeon of the United States, which lays only one, or at most two eggs, and is said to rear generally but one young one. Why is this bird so extraordinarily abundant, while others producing two or three times as many young are much less plentiful? The explanation is not difficult. The food most congenial to this species, and on which it thrives best, is abundantly distributed over a very extensive region, offering such differences of soil and climate, that in one part or another of the area the supply never fails. The bird is capable of a very rapid and long-continued flight, so that it can pass without fatigue over the whole of the district it inhabits, and as soon as the supply of food begins to fail in one place is able to discover a fresh feeding-ground. This example strikingly shows us that the procuring a constant supply of wholesome food is almost the sole condition requisite for ensuring the rapid increase of a given species, since neither the limited fecundity, nor the unrestrained attacks of birds of prey and of man are here sufficient to check it. In no other birds are these peculiar circumstances so strikingly combined. Either their food is more liable to failure, or they have not sufficient power of wing to search for it over an extensive area, or during some season of the year it becomes very scarce, and less wholesome substitutes have to be found; and thus, though more fertile in offspring, they can never increase beyond the supply of food in the least favourable seasons. Many birds can only exist by migrating, when their food becomes scarce, to regions possessing a milder, or at least a different climate, though, as these migrating birds are seldom excessively abundant, it is evident that the countries they visit are still deficient in a constant and abundant supply of wholesome food. Those whose organization does not permit them to migrate when their food becomes periodically scarce, can never attain a large population. This is probably the reason why woodpeckers are scarce with us, while in the tropics they are among the most abundant of solitary birds. Thus the house sparrow is more abundant than the redbreast, because its food is more constant and plentiful—seeds of grasses being preserved during the winter, and our farmyards and stubble-fields furnishing an almost inexhaustible supply. Why, as a general rule, are aquatic, and especially sea birds, very numerous in individuals? Not because they are more prolific than others, generally the contrary; but because their food never fails, the sea-shores and river-banks daily swarming with a fresh supply of small Mollusca and Crustacea. Exactly the same laws will apply to mammals. Wild cats are prolific and have few enemies; why then are they never as abundant as rabbits? The only intelligible answer is that their supply of food is more precarious. It appears evident, therefore, that so long as a country remains physically unchanged, the numbers of its animal population cannot materially increase. If one species does so, some others requiring the same kind of food must diminish in proportion. The numbers that die annually must be immense; and as the individual existence of each animal depends upon itself, those that die must be the weakest—the very young, the aged, and the diseased—while those that prolong their existence can only be the most perfect in health and vigour—those who are best able to obtain food regularly, and avoid their numerous enemies. It is, as we commenced by remarking, "a struggle for existence," in which the weakest and least perfectly organized must always succumb.

Now it is clear that what takes place among the individuals of a species must also occur among the several allied species of a group—viz., that those which are best adapted to obtain a regular supply of food, and to defend themselves against the attacks of their enemies and the vicissitudes of the seasons, must necessarily obtain and preserve a superiority in population; while those species which from some defect of power or organization are the least capable of counteracting the vicissitudes of food supply, etc., must diminish in numbers, and, in extreme cases, become altogether extinct. Between these extremes the species will present various degrees of capacity for ensuring the means of preserving life; and it is thus we account for the abundance or rarity of species. Our ignorance will generally prevent us from accurately tracing the effects to their causes; but could we become perfectly acquainted with the organization and habits of the various species of animals, and could we measure the capacity of each for performing the different acts necessary to its safety and existence under all the varying circumstances by which it is surrounded, we might be able even to calculate the proportionate abundance of individuals which is the necessary result.

If now we have succeeded in establishing these two points; first, *that the animal population of a country is generally stationary, being kept down by a periodical deficiency of food, and other checks;* and, secondly, *that the comparative abundance or scarcity of the individuals of the several species is entirely due to their organization and resulting habits, which, rendering it more difficult to procure a regular supply of food and to provide for their personal safety in some cases than in others, can only be balanced by a difference in the population which has to exist in a given area*—we shall be in a condition to proceed to the consideration of *varieties,* to which the preceding remarks have a direct and very important application.

Most or perhaps all the variations from the typical form of a species must have some definite effect, however slight, on the habits or capacities of the individuals. Even a change of colour might, by rendering them more or less distinguishable, affect their safety; a greater or less development of hair might modify their habits. More important changes, such as an increase in the power or dimensions of the limbs or any of the external organs, would more or less affect their mode of procuring food or the range of country which they inhabit. It is also evident that most changes would affect, either favourably or adversely, the powers of prolonging existence. An antelope with shorter or weaker legs must necessarily suffer more from the attacks of the feline carnivora; the passenger pigeon with less powerful wings would sooner or later be affected in its powers of procuring a regular supply of food; and in both cases the result must necessarily be a diminution of the population of the modified species. If, on the other hand, any species should produce a variety having slightly increased powers of preserving existence, that variety must inevitably in time acquire a superiority in numbers. These results must follow as surely as old age, intemperance, or scarcity of food produce an increased mortality. In both cases there may be many individual exceptions; but on the average the rule will invariably be found to hold good. All varieties will therefore fall into two classes—those which under the same conditions would never reach the population of the parent species, and those which would in time obtain and keep a numerical superiority. Now, let some alteration of physical conditions occur in the district—a long period of drought, a destruction of vegetation by locusts, the irruption of some new carnivorous animal seeking "pastures new"—any change in fact tending to render existence more difficult to the species in question, and tasking its utmost powers to avoid complete extermination; it is evident that, of all the individuals composing the species, those forming the least numerous and most feebly organized variety would suffer first, and, were the pressure severe, must soon become extinct. The same causes continuing in action, the parent species would next suffer, would gradually diminish in numbers, and with a recurrence of similar unfavourable conditions might also become extinct. The superior variety would then alone remain, and on a return to favourable circumstances would rapidly increase in numbers and occupy the place of the extinct species and variety.

The *variety* would now have replaced the *species*, of which it would be a more perfectly developed and more highly organized form. It would be in all respects better adapted to secure its safety, and to prolong its individual existence and that of the race. Such a variety *could not* return to the original form; for that form is an inferior one, and could never compete with it for existence. Granted, therefore, a "tendency" to reproduce the original type of the species, still the variety must ever remain preponderant in numbers, and under adverse physical conditions *again alone survive.* But this new, improved, and populous race might itself, in course of time, give rise to new varieties, exhibiting several diverging modifications of form, any of which, tending to increase the facilities for preserving existence, must, by the same general law, in their turn become predominant. Here, then, we have *progression and continued divergence* deduced from the general laws which regulate the existence of animals in a state of nature, and from the undisputed fact that varieties do frequently occur. It is not, however, contended that this result would be invariable; a change of physical conditions in the district might at times materially modify it, rendering the race which had been the most capable of supporting existence under the former conditions now the least so, and even causing the extinction of the newer and, for a time, superior race, while the old or parent species and its first inferior varieties continued to flourish. Variations in unimportant parts might also occur, having no perceptible effect on the life-preserving powers; and the varieties so furnished might run a course parallel with the parent species, either giving rise to further variations or returning to the former type. All we argue for is that certain varieties have a tendency to maintain their existence longer than the original species, and this tendency must make itself felt; for though the doctrine of chances or averages can never be trusted to on a limited scale, yet, if applied to high numbers, the results come nearer to what theory demands, and, as we approach to an infinity of examples, become strictly accurate. Now the scale on which nature works is so vast—the numbers of individuals and periods of time with which she deals approach so near to infinity, that any cause however slight, and however liable to be veiled and counteracted by accidental circumstances, must in the end produce its full legitimate results.

Let us now turn to domesticated animals, and inquire how varieties produced among them are affected by the principles here enunciated. The essential difference in the condition of wild and domestic animals is this, that among the former, their well-being and very existence depend upon the full exercise and healthy condition of all their senses and physical powers, whereas, among the latter, these are only partially exercised, and in some cases are absolutely unused. A wild animal has to search, and often to labour, for every mouthful of food—to exercise sight, hearing, and smell in seeking it, and in avoiding dangers, in procuring shelter from the inclemency of the seasons, and in providing for the subsistence and safety of its offspring. There is no muscle of its body that is not called into daily and hourly activity; there is no sense or faculty that is not strengthened by continual exercise. The domestic animal, on the other hand, has food provided for it, is sheltered, and often confined, to guard it against the vicissitudes of the seasons, is carefully secured from the attacks of its natural enemies, and seldom even rears its young without human assistance. Half of its senses and faculties are quite useless; and the other half are but occasionally called into feeble exercise, while even its muscular system is only irregularly called into action.

Now when a variety of such an animal occurs, having increased power or capacity in any organ or sense, such increase is totally useless, is never called into action, and may even exist without the animal ever becoming aware of it. In the wild animal, on the contrary, all its faculties and powers being brought into full action for the necessities of existence, any increase becomes immediately available, is strengthened by exercise, and must even slightly modify the food, the habits, and the whole economy of the race. It creates as it were a new animal, one of superior powers, and which will necessarily increase in numbers and outlive those inferior to it.

Again, in the domesticated animal all variations have an equal chance of continuance; and those which would decidedly render a wild animal unable to compete with its fellows and continue its existence are no disadvantage whatever in a state of domesticity. Our quickly fattening pigs, short-legged sheep, pouter pigeons, and poodle dogs could never have come into existence in a state of nature, because the very first step towards such inferior forms would have led to the rapid extinction of the race; still less could they now exist in competition with their wild allies. The great speed but slight endurance of the race horse, the unwieldy strength of the ploughman's team, would both be useless in a state of nature. If turned wild on the pampas, such animals would probably soon become extinct, or under favourable circumstances might each lose those extreme qualities which would never be called into action, and in a few generations would revert to a common type, which must be that in which the various powers and faculties are so proportioned to each other as to be best adapted to procure food and secure safety—that in which by the full exercise of every part of his organization the animal can alone continue to live. Domestic varieties, when turned wild, *must* return to something near the type of the original wild stock, *or become altogether extinct.*

We see, then, that no inferences as to varieties in a state of nature can be deduced from the observation of those occurring among domestic animals. The two are so much opposed to each other in every circumstance of their existence, that what applies to the one is almost sure not to apply to the other. Domestic animals are abnormal, irregular, artificial; they are subject to varieties which never occur and never can occur in a state of nature: their very existence depends altogether on human care; so far are many of them removed from that just proportion of faculties, that true balance of organization, by means of which alone an animal left to its own resources can preserve its existence and continue its race.

The hypothesis of Lamarck—that progressive changes in species have been produced by the attempts of animals to increase the development of their own organs, and thus modify their structure and habits—has been repeatedly and easily refuted by all writers on the subject of varieties and species, and it seems to have been considered that when this was done the whole question has been finally settled; but the view here developed renders such an hypothesis quite unnecessary, by showing that similar results must be produced by the action of principles constantly at work in nature. The powerful retractile talons of the falcon and the cat tribes have not been produced or increased by the volition of those animals; but among the different varieties which occurred in the earlier and less highly organized forms of these groups, *those always survived longest which had the greatest facilities for seizing their prey.* Neither did the giraffe acquire its long neck by desiring to reach the foliage of the more lofty shrubs, and constantly stretching its neck for the purpose, but because any varieties which occurred among its antetypes with a longer neck than usual *at once secured a fresh range of pasture over the same ground as their shorter-necked companions, and on the first scarcity of food were thereby enabled to outlive them.* Even the peculiar colours of many animals, especially insects, so closely resembling the soil or the leaves or the trunks on which they habitually reside, are explained on the same principle; for though in the course of ages varieties of many tints may have occurred, *yet those races having colours best adapted to concealment from their enemies would inevitably survive the longest.* We have also here an acting cause to account for that balance so often observed in nature—a deficiency in one set of organs always being compensated by an increased development of some others—powerful wings accompanying weak feet, or great velocity making up for the absence of defensive weapons; for it has been shown that all varieties in which an unbalanced deficiency occurred could not long continue their existence. The action of this principle is exactly like that of the centrifugal governor of the steam engine, which checks and corrects any irregularities almost before they become evident; and in like manner no unbalanced deficiency in the animal kingdom can ever reach any conspicuous magnitude, because it would make itself felt at the very first step, by rendering existence difficult and extinction almost sure soon to follow. An origin such as is here advocated will also agree with the peculiar character of the modifications of form and structure which obtain in organized beings—the many lines of divergence from a central type, the increasing efficiency and power of a particular organ through a succession of allied species, and the remarkable persistence of unimportant parts such as colour, texture of plumage and hair, form of horns or crests, through a series of species differing considerably in more essential characters. It also furnishes us with a reason for that "more specialized structure" which professor Owen states to be a characteristic of recent compared with extinct forms, and which would evidently be the result of the progressive modification of any organ applied to a special purpose in the animal economy.

We believe we have now shown that there is a tendency in nature to the continued progression of certain classes of *varieties* further and further from the original type—a progression to which there appears no reason to assign any definite limits—and that the same principle which produces this result in a state of nature will also explain why domestic varieties have a tendency to revert to the original type. This progression, by minute steps, in various directions, but always checked and balanced by the necessary conditions, subject to which alone existence can be preserved, may, it is believed, be followed out so as to agree with all the phenomena presented by organized beings, their extinction and succession in past ages, and all the extraordinary modifications of form, instinct, and habits which they exhibit.

~ 4 ~

The World

This chapter spans more than fifty years of Wallace's life, from his return to England from the Malay Archipelago in 1862 to his death in 1913. The breadth of his writings and my preference for providing a sample of his range make this chapter less thematically unified than the others. The first two selections deal with Wallace's views on human evolution. The first reading is a short selection from one of the earliest of many essays he wrote on this topic and is of particular historical interest because it is the first time that he invoked in print a higher power to account for the origin of humans. The second reading, Wallace's essay "Evolution and Character," jumps way ahead to 1912 and is one of the most synthetic chapters he wrote, bringing together his overarching views on evolution and human society. Although neither of these two readings is directly "from the field," his views on humans were strongly influenced by the twelve years he lived in the tropics among native peoples as well as by his observations of Western culture. The broader context for the development of his ideas on human evolution is described below. The next reading stands by itself as a summary of his impressions of North America, written in 1887 at the end of his ten-month trip. The last selection comprises a few remembrances written by his son and daughter.

While unpacking his many crates and settling back into life in London after eight years in the Malay Archipelago, Wallace once again began attending meetings at the Zoological, Entomological, and Linnean Societies, re-entering the scientific arena. In addition to presenting numerous papers on insects and birds, he turned his attention to humans, interpreting patterns of distribution and variation according to the theory of natural selection. In writing his observations on the Malayan and Papuan peoples, he became involved in highly controversial questions about human evolution.

The combination of the publication of *On the Origin of Species* in 1859 with reports that same year of recent archeological finds of stone tools and human fossils led to an explosion of debate about human origins.[1] Did human beings evolve from apes? Are human races varieties of *Homo sapiens,* or are they separate species? Precisely when did humans appear in the history of Earth? Were early humans the same species as we are? How do we define what it means to be human? As a new consensus emerged which greatly extended the time scale of human evolution, questions about man's origins became more focused than ever before. Once the purview of archeology, the new evidence and theories of human evolution were now fair game for geologists, anthropologists, and naturalists. Victorian debates about man, nature, God, and society involved not only scientists, but theologians, philosophers, and social critics as well.

With a series of papers in the 1860s Wallace entered the fray over man's place in nature in which he made complex and sometimes contradictory arguments about the evolution of human beings. In 1864 Wallace published his first two papers on man in which he relied on evolution by natural selection to explain the development of human races.[2] The more general of the two papers addressed the hotly debated question of whether the races of man are local variations or permanent, distinct species. Here he argued that humankind, a single species, had differentiated into separate races early in its evolution. He posited that natural selection had acted on man's higher faculties—speech, weapon making, division of labor, anticipation of the future, restraint of the appetites, moral, social, and sympathetic feelings—because these, not physical

Fig. 21. Photograph of Alfred R. Wallace in 1869, at age
forty-six, taken by his brother-in-law Thomas Sims, a
professional photographer. From *My Life,* 1: facing p. 386.

adaptations, would have been the dominant influences on his survival. As
much as Darwin liked this paper for its clear articulation of natural selection acting on man's intellect, he would soon come to disagree with Wallace's claim that forces other than natural selection are required to account for human evolution.

By 1869 (Figure 21), Wallace had come to believe that natural selection could not account for the origin of man's large brain and for what he
referred to as man's "higher mental capacities." Instead, Wallace spoke of
some unknown higher law, a superior intelligence, a spiritual influx that
makes matter, life, consciousness (by which he meant animal sensation),
and intellectual life possible. This "mystical turn," so to speak, made its

first appearance in a review Wallace wrote of two geology books written by his good friend, the renowned geologist Sir Charles Lyell. Wallace's review could have simply celebrated the fact that Lyell had finally come around to accepting the theory of evolution by natural selection. Indeed, Wallace did elaborate on the significance of Lyell's about-face on evolution, but he also used the review to champion two other concerns: his fascination with a speculative theory of astronomical causes of climatic change, and human evolution.

In the portion of the review selected here, he explained his reason for believing that natural selection could not account for the origin of those characteristics that set humankind above other forms of life. His argument was strictly "Darwinian" in that he could not see how any trait, physical or mental, could be the product of natural selection if it was not useful in the struggle for existence.[3] In this first of many discussions of the limits of natural selection, he proposed that several physical traits, such as hairless skin and erect posture, were the product of a guiding intelligence, as were the higher moral and emotional capacities. Beginning in 1870 and continuing for the rest of his life, he expanded these ideas in several long essays, modifying his views to the point where he saw only mental traits, not physical ones, as inexplicable by the theory of natural selection; he held firm in his belief in a purposeful universe guided by a superior mind.

Wallace's formulation of a more spiritual and less naturalistic perspective on evolution took shape during the years in which he became entranced with the study of psychic phenomena. In the mid-1860s he began attending seances, and this led him to investigate a whole range of so-called supernatural phenomena, including communication with the dead, the action of magnets and crystals on people's minds, mesmeric trances, crystal seeing, and spirit photographs. Wallace's fascination with mental phenomena dated to his youth, when, at the age of twenty-one, he heard a lecture on mesmerism. He had performed his own experiments on students and other willing subjects, including Indians he met in Brazil. The positive results of these experiments, along with his experiences in London in the 1860s, convinced him that psychic phenomena were genuine and led him to disagree outspokenly with those who, without observation and study, denied the possibility of such phenomena.[4]

It is important to realize that Wallace's willingness to investigate the supernatural was part of an earnest, intellectual inquiry into all of nature. It is probably misleading to say, as many historians have, that he converted from science to spiritualism; rather, he struggled with several difficult questions about the limits of natural selection.[5] After all, natural selection, as it was understood then as well as now, acts on existing variations. It does not explain the origin of variations, much less those as complex as consciousness and moral sentiments. Focused as Wallace was on the notion of utility, he could not fathom how musical and artistic talent, mathematical genius, wit, or intellectual endeavors could possibly have been useful in the struggle for existence, but he was aware of two potential explanations: that certain moral capacities were in fact useful to the survival of a social being and that some traits may be by-products of other changes that have been selected for. He argued at length that the nonutilitarian nature of the higher mental and moral capacities, their origin, and their tremendous variability were inconsistent with the laws of natural selection.

The second reading is a mature version of his later ideas on evolution writ large, from the origins of matter to the development of humans. For Wallace, eventually, the whole of nature, from the inorganic to the organic to the spiritual, had developed by means of a unified evolutionary process driven by a single purpose: to evolve spiritual beings capable of indefinite life and perfectibility. In a sense, Wallace continued a centuries-old line of reasoning: that the ultimate purpose of nature and of the struggle between good and evil was to enhance human development. In *The World of Life*, published in 1910, Wallace, then well into his eighties, amassed evidence for this progressionist world view. Yet in the essay selected for inclusion here, a chapter he contributed to a book called *Character and Life* (1912), we find some refreshing results of this seemingly old-fashioned view of the perfectibility of the human spirit combined with his social and political views. In his many encounters with "savages," he had found them to show a higher level of morality, to have more respect for others, and to live in communities with greater social justice than people living in supposedly more advanced cultures. One of the most elegant passages expressing Wallace's warnings about European chauvinism in relation to native cultures is found in the concluding pages of *The Malay Archipelago*. In the essay printed here, he went even further, claiming that not only was modern man no better than so-called savages, but that man had not advanced morally or intellectually since the earliest phases of human evolution. Examples of Wallace's irreverence abound in this chapter. He lamented, for instance, that widespread and costly religious and educational agencies have not improved human character a whit and that sympathy with children is not even a condition of entering the teaching profession. But he held great hopes for the future based on a truly moral education in conjunction with selection through marriage. "Only when a greatly improved social system renders all our women economically and socially free to choose" and a complete education will have taught them the importance of their choice, then, Wallace believed, the higher sorts of men will be selected, and the human race will thereby be permanently improved.

𝒦 His commitment to an open-minded analysis of hypotheses of all kinds served Wallace well in pursuing scientific as well as so-called supernatural interests. Whether it was animal instincts or protective coloration, changes in climate and their effect on plant distribution, or the importance of patterns of geographical distribution for evolutionary history, he continued to read avidly, refine his ideas, and publish. At the same time he also addressed political and sociological issues, from questions about education and museum design to problems of land reform and social justice.

The same breadth is evident in the tour he made of North America in 1886–87. The tour was built around an invitation to deliver a series of lectures at the prestigious Lowell Institute in Boston. These lectures, as well as those he presented as he visited universities in the United States and Canada, were grounded in his understanding of organic evolution. The lectures covered six main topics under the rubric of "The Darwinian Theory," including colors of animals and plants, mimicry, and the physical and biological relations of islands and continents. Well-received as an eminent man of science, Wallace enjoyed the opportunities for extended discussions with American scientists, as well as visits to museums, libraries, and other learned academies. But his trip was not limited to academic lecturing. He also gave talks to and visited with people whose interests focused on political and sociological debates and with many people he described as spiritualists.

He spent about five months in the eastern United States, using Boston and then Washington, D.C., as a base from which to travel. At one formal dinner at which many of Boston's most celebrated men of letters were gathered, Wallace found that he "was not so much impressed by the Boston celebrities as I ought to have been."[6] But this was an exception to his many productive and amicable exchanges with a great variety of intellectual, literary, and scientific people. He then traveled by train from Washington, D.C., to San Francisco, stopping along the way to lecture and explore the country. Observant as ever, on the lookout for rare plants and for evidence of former ice ages, Wallace took advantage of numerous stops on his rail journey to get a feeling for the various landscapes as he crossed the continent.

In San Francisco he was able to spend time with his brother John and his family (John had emigrated to California in 1849), making several excursions with him to explore Yosemite and other natural sites. Wallace accepted an invitation to visit the home of Senator Leland Stanford, a fellow spiritualist whom he had met in Washington. Wallace enjoyed visiting the site of the future Stanford University but found the millionaire's capitalist viewpoint fundamentally flawed. Perhaps the most remarkable of his California experiences was his highly successful lecture on spiritualism. As Wallace noted in his autobiography, "the audience was most attentive, and it was not only a better audience, but the net proceeds were more than for any single scientific lecture I gave in America."[7] Wallace set aside two weeks to explore the alpine plants of the Sierras and the Rockies, after which he traveled by rail to Chicago and then Montreal, Quebec, and home via steamer to Liverpool.

The reading on Wallace's North American tour was originally written as a concluding entry in the journal he kept during his trip. It may surprise or dismay some readers to find how strongly negative were his impressions of the land and people in the United States. The depth of Wallace's commitment to the fair use of land and his sense of social justice are nowhere expressed more pointedly than in this essay. Although reticent in his personal relations, he was passionate and unrestrained in his conviction that the well-being of society, as part of nature, depends on a political economy based in socialism.

✍ Although not without its difficulties, Wallace's old age was enormously productive. He continued to read with great appetite, to write letters and books, and to lecture. In 1902, at the age of seventy-nine, he

Fig. 22. Wallace's home, Old Orchard, at Broadstone.
From *My Life*, 2: frontispiece.

built a new house and enjoyed cultivating his new, expansive garden. His daughter, a teacher, came to live with her parents in 1903, and several of her pupils shared the Wallace home at "Old Orchard" as well (see Figure 22). Wallace stayed in close contact with his son, who edited his father's later books. His immediate family members were all present when Wallace died at home in 1913. Perhaps the unity that Wallace saw in the evolution of life was, in part, a projection of the contentment and harmony of his personal life.

The remembrances of Wallace's son and daughter, William and Violet, originally appeared in a book of letters and reminiscences published by Wallace's friend James Marchant shortly after Wallace died. A portion of their chapter is included here to provide a down-to-earth portrait of Wallace. The descriptions by his children of how he looked and walked, how he spent his time at home, the books he read for pleasure, and what sorts of vacations the family took give us a firsthand glimpse of what he was like. We also find something that Wallace himself did not leave us—a sense of the presence of his wife Annie, whose love of plants was shared with her family. She was twenty-three years younger than her husband, lived with him for forty-seven, and after suffering from a long illness, died within a year of his death.

`~ ~`

Limits of Natural Selection in Human Evolution

In adopting the views of Mr. Darwin, Sir Charles Lyell carries them out to their legitimate results, and does not shrink from the logical necessity, of the derivation of man from the lower animals; and he has written a very interesting chapter on the "Origin and Distribution of Man." Into this subject, however, we cannot now enter, except to remark briefly on some aspects of the question which all who have hitherto written upon it seem to have neglected.

It would certainly appear in the highest degree improbable, that the whole animal kingdom, from the lowest zoöphytes up to the horse, the dog, and the ape, should have been developed by the simple action of natural laws, and that the animal man, so absolutely identical with them in all the main features and many of the details of his organization, should have been formed in some quite other unknown way. But if the researches of geologists and the investigations of anatomists should ever demonstrate that he was derived from the lower animals in the same way that they have been derived from each other, we shall not be thereby debarred from believing or from proving that his intellectual capacities and his moral nature were not wholly developed by the same process. Neither natural selection nor the more general theory of evolution can give any account whatever of the origin of sensational or conscious life. They may teach us how, by chemical, electrical, or higher natural laws, the organized body can be built up, can grow, can reproduce its like; but those laws and that growth cannot even be conceived as endowing the newly-arranged atoms with consciousness. But the moral and higher intellectual nature of man is as unique a phenomenon as was conscious life on its first appearance in the world, and the one is almost as difficult to conceive as originating by any law of evo-

"Geological Climates and the Origin of Species," *Quarterly Review* 126 (1869):391–94.

lution as the other. We may even go further, and maintain that there are certain purely physical characteristics of the human race which are not explicable on the theory of variation and survival of the fittest. The brain, the organs of speech, the hand, and the external form of man, offer some special difficulties in this respect, to which we will briefly direct attention.

In the brain of the lowest savages, and, as far as we yet know, of the prehistoric races, we have an organ so little inferior in size and complexity to that of the highest types (such as the average European), that we must believe it capable, under a similar process of gradual development during the space of two or three thousand years, of producing equal average results. But the mental requirements of the lowest savages, such as the Australians or the Andaman islanders, are very little above those of many animals. The higher moral faculties and those of pure intellect and refined emotion are useless to them, are rarely if ever manifested, and have no relation to their wants, desires, or well-being. How, then, was an organ developed so far beyond the needs of its possessor? Natural selection could only have endowed the savage with a brain a little superior to that of an ape, whereas he actually possesses one but very little inferior to that of the average members of our learned societies.

Again, what a wonderful organ is the hand of man;* of what marvels of delicacy is it capable, and how greatly it assists in his education and mental development! The whole circle of the arts and sciences are ultimately dependent on our possession of this organ, without which we could hardly have become truly human. This hand is equally perfect in the lowest savage, but he has no need for so fine an instrument, and can no more fully utilise it than he could use without instruction a complete set of joiner's tools. But, stranger still, this marvellous instrument was foreshadowed and prepared in the Quadrumana; and any person who will watch how one of these animals uses its hands, will at once perceive that it possesses an organ far beyond its needs. The separate fingers and the thumb are never fully utilised, and objects are grasped so clumsily, as to show that a much less specialised organ of prehension would have served its purpose quite as well; and if this be so, it could never have been produced through the agency of natural selection alone.

* See the admirable volume on the "Hand," by the late Sir Charles Bell, in the Bridgewater Treatise.

We have further to ask—How did man acquire his erect posture, his delicate yet expressive features, the marvellous beauty and symmetry of his whole external form—a form which stands alone, in many respects more distinct from that of all the higher animals than they are from each other? Those who have lived much among savages know that even the lowest races of mankind, if healthy and well-fed, exhibit the human form in its complete symmetry and perfection. They all have the soft, smooth skin absolutely free from any hairy covering on the dorsal line, where all other mammalia from the Marsupials up to the Anthropoid apes have it most densely and strongly developed. What use can we conceive to have been derived from this exquisite beauty and symmetry, and this smooth, bare skin, both so very widely removed from his nearest allies? And if these modifications were of no physical use to him—or if, as appears almost certain in the case of the naked skin, they were at first a positive disadvantage—we know that they could not have been produced by natural selection. Yet we can well understand that both these characters were essential to the proper development of the perfect human being. The supreme beauty of our form and countenance has probably been the source of all our æsthetic ideas and emotions, which could hardly have arisen had we retained the shape and features of an erect gorilla; and our naked skin, necessitating the use of clothing, has at once stimulated our intellect, and by developing the feeling of personal modesty may have profoundly affected our moral nature.

The same line of argument may be used in connexion with the structural and mental organs of human speech, since that faculty can hardly have been physically useful to the lowest class of savages; and if not, the delicate arrangements of nerves and muscles for its production could not have been developed and co-ordinated by natural selection. This view is supported by the fact that, among the lowest savages with the least copious vocabularies, the capacity of uttering a variety of distinct articulate sounds, and of applying to them an almost infinite amount of modulation and inflection, is not in any way inferior to that of the higher races. An instrument has been developed in advance of the needs of its possessor.

This subject is a vast one, and would require volumes for its proper elucidation, but enough, we think, has now been said, to indicate the possibility of a new stand-point for those who cannot accept the theory of evolution as expressing the whole truth in regard to the origin of man. While admitting to the full extent the agency of the same great laws of organic development in the origin of the human race as in the origin of all organized beings, there yet seems to be evidence of a Power which has guided the action of those laws in definite directions and for special ends. And so far from this view being out of harmony with the teachings of science, it has a striking analogy with what is now taking place in the world, and is thus strictly uniformitarian in character. Man himself guides and modifies nature for special ends. The laws of evolution alone would perhaps never have produced a grain so well adapted to his uses as wheat; such fruits as the seedless banana, and the breadfruit; such animals as the Guernsey milch-cow, or the London dray-horse. Yet these so closely resemble the unaided productions of nature, that we may well imagine a being who had mastered the laws of development of organic forms through past ages, refusing to believe that any new power had been concerned in their production, and scornfully rejecting the theory that in these few cases a distinct intelligence had directed the action of the laws of variation, multiplication, and survival, for his own purposes. We know, however, that this has been done; and we must therefore admit the possibility, that in the development of the human race, a Higher Intelligence has guided the same laws for nobler ends.

Such, we believe, is the direction in which we shall find the true reconciliation of Science with Theology on this most momentous problem. Let us fearlessly admit that the mind of man (itself the living proof of a supreme mind) is able to trace and to a considerable extent has traced, the laws by means of which the organic no less than the inorganic world has been developed. But let us not shut our eyes to the evidence that an Overruling Intelligence has watched over the action of those laws so directing variations and so determining the accumulation, as finally to produce an organization sufficiently perfect to admit of, and even to aid in, the indefinite advancement of our mental and moral nature.

~❧ ❧~

Spiritualism and Human Evolution

In the very brief discussion of a great subject here attempted, I limit myself to a strict application of the modern doctrine of Organic Evolution to

"Evolution and Character," in *Character and Life,* edited by Percy L. Parker, 1912; first published in *Fortnightly Review* 83 (1908):1–24.

certain definite inquiries as to the probable development of human faculty or character. In doing so I first endeavour to define what is meant by evolution in general and by organic evolution in particular, and then proceed to show what are the essential agencies or processes by means of which the latter carries on its work. Dealing next with our more special subject, I inquire into the supposed differences between the minds of savages and those of civilised men, and also into those between our human nature to-day and that which existed in the earliest historic or pre-historic ages of which we have any records. Having thus arrived at certain probable conclusions, I proceed to suggest the conditions and agencies which are alone adequate to bring about a continuous advance in the average character of man during future ages.

The term *Evolution*, though now so generally used, is yet often misunderstood. It is supposed by many, perhaps by the majority of non-scientific persons, to indicate a great scientific theory which is applicable to and explains all the phenomena of the universe. But this is, very largely, an erroneous view. It is true that by many of its advocates it is held to be universally applicable, yet it has, so far, only given us a fairly complete explanation in certain departments of nature, and even in these it never carries us back to the beginnings of things; while over some of the broadest fields of scientific research it has been almost entirely inoperative.

Its essential idea is that of the continuity of all the phenomena of nature—that everything we see on the earth or in the spaces around us is not permanent, but has arisen out of something that preceded it. It is thus opposed to the old, and to some extent still prevalent, idea of creation—that things as we now see them have existed from some remote but definite epoch when they came into existence by the act or fiat of a supreme power—the great First Cause. Evolution is thus a general statement that everything is, and always has been, slowly changing under the influences of natural laws and processes; but, except in a few cases, it cannot give a precise account of the methods and causes of the changes, still less can it carry us back to any beginning of the universe. It thus formulates a general process, but is unable to give us any full explanation of that process.

Although the philosophers of Greece had vague ideas of evolution, which were elaborately worked out by the Roman poet Lucretius in his great poem "On the Nature of Things," yet their views obtained no general acceptance until our own era, owing mainly to the positive statements as to the creation of the universe in the "Old Testament," and the very general acceptance of that record as the Word of God. Even the very surface of the earth was held to be unchangeable, as implied in the term "the everlasting hills," while less than fifty years ago so great a writer as George Borrow could speak of a waterfall as being in all details as it was "since the day of creation, and will probably remain to the day of judgment."

The modern view of continuous change by natural forces was first applied to the surface and structure of the earth, by more or less careful observation of the facts and phenomena included in the modern science of geology. It began with a few acute observers in the seventeenth century, among whom were Leibnitz and Hooke, followed by numerous Italian writers in the eighteenth century, together with the Germans, Pallas and Werner; while our own countryman, Hutton, for the first time laid down the great principle of modern earth-study, that we can only understand the past by a careful study of the various changes we now perceive to be in progress. This great principle was afterwards most skilfully applied by Sir Charles Lyell, who devoted a long life to the production and continuous amplification of his monumental work, *The Principles of Geology*. The idea of evolution was thus applied in detail to one of the greatest and most complex departments of human knowledge.

In the two other great sciences dealing with the constitution of the inorganic world—chemistry and astronomy—progress was for a long time necessarily limited to the study of facts and phenomena, with the laws to which those phenomena are immediately due; and the conclusions arrived at pointed rather to stability and permanence than to that progressive and unceasing change that is the keynote of evolution. The great mathematicians, who at the beginning of the nineteenth century worked at the motions and disturbances in our solar system in accordance with the law of gravitation, came to the conclusion that the system was a stable one, that all irregularities were slight and temporary, and that the planets and their satellites were so arranged that their present positions and motions might continue for ever without any destructive changes.

In like manner it was for long a fundamental doctrine of chemistry that the elements were fixed and unchangeable, and the belief of the alchemists that other substances might be converted into gold was held to be as baseless and as unscientific a dream as the idea that matter itself was destructible.

But in our own day, and largely by the work of men still living, all these assumptions of indestructibility and permanence have been rudely shaken or altogether given up as in their turn unscientific. Through the development of what may be well termed the two modern sciences—electricity and spectrum analysis—together with the systematic study of the long-neglected phenomena of meteors and meteor systems, quite new conceptions have been reached as to the constitution of the universe, ascending, on the one hand, to the nature and origin of the myriad stars and suns and nebulæ which constitute our universe, and descending on the other to the nature, the constitution, and even the instability of matter itself, as indicated by the strange and almost incredible phenomena presented by radio-active substances.

By these various advances in many directions we have attained the certainty that the great principle of evolution pervades the entire realm of nature, from the faint specks of star-dust on the farthest limits of our stellar universe, down to what were once supposed to be the indestructible atoms of matter itself, now believed to be complex systems of electric force-points, subject to disturbances and even to absolute disintegration. We thus seem able dimly to comprehend on the one hand the evolution of matter itself, with its marvellous properties, which enable it to become manifest in the myriad forms made known to us by the chemist or existing in the vast laboratory of nature; on the other hand, the evolution of this matter into the inconceivably vast and complex stellar universe. Everywhere we behold a state of flux, of development, and also, apparently, of decay. Every increase of knowledge seems to imply that the material universe is a vast organism which must have had a beginning and will have an end—which was born and will die. The dissipation of energy and the disintegration of matter alike render this conclusion logically certain.

The preceding remarks apply only to what may be termed inorganic or physical evolution, which necessarily preceded and prepared the way for the evolution of the organic world—an evolution which is utterly unlike anything which preceded it and which has culminated in the production of man—the one being who is able, to some extent, to comprehend the universe of which he forms a part, to penetrate to its remotest confines, to study its laws and to speculate on its nature, its origin, and its destiny. Hence we may naturally conclude that the final law and purpose of the whole universe was the development of so marvellous a being, who has been deemed to be "a little lower than the angels" and "in apprehension like a god."*

It will be well to note here the fundamental difference between organic and inorganic evolution, a difference so great and so radical that it is somewhat misleading to use the same term to describe them both. The changes that occur in the inorganic matter of the universe are of three kinds: (1) The changes of external form, and to some extent of internal structure, caused by the disintegration and aggregation of masses of matter, as in the formation of most rocks and in the successive modifications of the earth's surface. (2) The changes in the state of matter—solid, liquid, and gaseous— dependent on the amount of heat received from without or generated within cosmical bodies. (3) Chemical changes, by which the very constitution of matter becomes so modified as to give each new compound a special character and properties which it usually retains unchanged for indefinite periods. The various kinds of matter produced by these changes seem to be permanent so long as the conditions remain the same, but, except in the case of crystals, they are of no definite shape or size. By changes of conditions they become to some extent interchangeable, but, except when subjected to these changes, they remain inert, or subject to excessively slow processes of degradation or decomposition, whence the common term "dead matter."

Organic evolution, on the other hand, leads to the production of highly complex individual entities of definite forms, each in a state of constant internal movement, each permeated by liquids and gases by means of which they assimilate new matter from the outer world, change that matter into new forms that occur nowhere else in nature, and are enabled to carry on the mysterious processes of growth and reproduction. Each of these individuals, beginning with minute cells in the body of a parent, passes through a regular cycle of growth, maturity, and decay, culminating in what we term death, when all its regular internal motions cease, it becomes disintegrated by the agency of lower organisms, and finally helps to build up other forms of life. Each reproduces its kind almost identical in all respects with the parent, thus forming a cycle which was long believed to be perpetual and unchanging, the species of animals and plants being held to be fixed enti-

* In the author's work, *Man's Place in the Universe*, the various lines of evidence leading to this conclusion have been fully set forth.

ties produced by some act of creative power. The great and unique phenomenon of the organic world is reproductive growth by the absorption and transformation of inorganic and organic matter, and the building up again and again of a highly complex organism from a single cell. It is this wonderful process that we term Life, whether manifested in animals which possess sensation and in their higher forms consciousness, or in plants which there is no reason to believe possess these faculties.

At about the middle of the last century so great a man of science as Sir John Herschel spoke of the mode of origin of the various species of animals and plants as being the "mystery of mysteries"; for although many writers had discarded "special creation" and had expressed a belief in evolution through the normal process of generation, yet no one had shown *how* the various species and genera had been produced, or by what means the wonderful adaptation of each to its special conditions of existence had been brought about. The problem was, however, solved by Charles Darwin almost coincidently with the other great advances in the domain of inorganic nature already alluded to.

The fundamental law which he discovered, and was the first to develop in all its far-reaching results, is that of "Natural Selection," or the "Survival of the Fittest." This great law of nature is the result of a group of well-known and universal facts—(1) the enormous powers of increase of all organisms, an increase so great that any one of them, if left alone in an unoccupied continent, would fill it to overflowing in a few years or centuries. As, however, the whole earth is already occupied, this simultaneous increase of all the thousands of species in every country produces a "struggle for existence," there being no room for the new-comers under average conditions till the parents have ceased to exist; and as all the higher animals (and plants) live several years, breeding every year, it is evident that on the average all their progeny must die from various causes before, or shortly after, they arrive at maturity except *one pair* to replace the two parents.

Now comes the question, is the destruction of the superfluous thousands a matter of pure chance, or is there any cause why certain individuals should survive the rest? If the offspring were always identical copies of the parents, not only in external form, but in every internal character and quality, in health, in acuteness of the senses, in activity, and in all the mental powers and faculties, then we should be obliged to impute to chance alone the destruction of ninety-nine while one survived. But we know there is no such similarity. In every large family of children considerable diversities occur as the rule rather than as the exception. In every litter of kittens or of puppies there are similar differences; while it has been through the selection of some of these varieties and the rejection of others that all our fine breeds of horses, cattle, and sheep have been produced, as well as all our fancy pigeons and poultry. And we now know that exactly the same thing occurs among plants as well as animals. By collecting and examining hundreds or thousands of individuals in one district and during one season, an amount of variation is found to exist much greater than anything that even Darwin expected. He sometimes spoke of nature having to wait for "favourable variations"; of natural selection being powerless unless "favourable variations" occurred when required. But these doubts and hesitations are utterly needless. There are *always* favourable variations in every direction, and in ample quantity. Take any measurable character you please, and in 50 or 100 or 1,000 individuals about one-third or one-fourth will be considerably above or below the mean, so much so as to be distinctly visible to us, while about one-third or one-half are so near the average of the whole that on a cursory inspection we should say they were all alike. But as, every year, only from one-tenth to one-hundredth of the young of a given species can survive, there is always an ample supply of "favourable variations." We must remember also that nature does not select, as we are often obliged to do, by the size or form of any limb, part, or organ, but by the resultant *qualities*, and we know that these qualities vary as much as the parts of the body we can measure or estimate. Horses, from the same or equally good parents, vary in speed and in endurance; dogs in acuteness of smell or of sight; sheep in the thickness of their wool; cows in their milk-giving capacity. In all these cases there is a considerable amount of better or of worse, offering ample "favourable variations." During an exceptionally severe winter only the swiftest and most enduring wolves survive, the rest perishing of cold and hunger. In a prolonged drought it is only the tallest giraffes that find food enough to support life; and thus, by a periodical weeding-out of all but the very best—the fittest to survive under these unfavourable conditions—the standard of efficiency in each species is preserved by the rigid destruction of the less fit. It must always be remembered that, although the *average* population of each species varies very little during long periods, yet there may be considerable fluctuations annually. Some seasons will favour one species, some another; we then notice the abundance of certain birds or insects, generally followed a year or two later by a corresponding scarcity, keeping up the balance of the various forms of life in generally uniform proportions so long as the natural conditions, or "environment," continue to be the same or nearly the same.

To anyone who has thoroughly grasped the extent and universality of variation within the limits of every common or widely-spread species, it will be at once evident that the very same causes which preserve each species in exact adjustment to its environment, will also, when that environment changes in any direction, enable it to become automatically adjusted to the new conditions. This must be the case, because all alterations in environment are necessarily very gradual. Changes of climate require thousands of years before they attain an amount greater than occurs during the ordinary seasonal or periodical changes to which all animals and plants are already adjusted. If a new enemy enters a country, it requires a number of years, perhaps centuries, before it can become itself adapted to the new conditions and increase to an extent sufficient to endanger any considerable number of other species. Some of the weaker kinds, and a proportion of the young of the stronger and more numerous, will no doubt fall a prey; but this will itself lead to the adaptation and improvement of the remainder, since those that escape will inevitably be those who are best fitted, either by swiftness, or strength, or cunning, or by possessing some special coloration or peculiar habit that conceals them from the enemy. The very danger itself leads to such a gradual modification of the sufferers as to enable those that escape to become the progenitors of a race better fitted to cope with the new danger.

Even more certainly automatic are the effects of climatic or food changes. To some these changes will be injurious, to some indifferent, to a few perhaps beneficial. The former will be weakened and thus fall a prey to the other causes of destruction, while the latter, being actually improved in health, will always furnish a large proportion of those that survive and become the parents of new races.

It is thus evident that those species which are already best adapted to the environment in a large country, that have a wide geographical range and exist in very great numbers, will be those that, by furnishing a large amount of variation every year, will be best able to become rapidly modified into an exact adaptation to any new conditions that may arise. These were termed by Darwin dominant species; and, owing to their general variability and enormous numbers, it is these that usually become the parents of a whole series of diverging species when exposed to a variety of new conditions in different parts of their area.

But besides these dominant forms other species exist which are rare or local, sometimes inhabiting only a very limited area, or being specially adapted to a very restricted mode of life—as when a caterpillar feeds on one species of plant only, and that not a common one. These are the species which are either dying out from want of power to compete with other species, or are so specially adapted to a limited environment that a comparatively slight change leads to their extinction. It is these which become extinct without leaving modified descendants, as seen in the countless genera and families which successively died out in each geological epoch. The dominant groups, on the other hand, can often be traced far back in geological time, as in the well-known case of the horse tribe, the cats, and some others; or they are those which, though of comparatively recent origin—as the deer and the antelopes—are so well adapted to existing conditions that they have spread over wide areas of the earth's surface with numerous specific forms adapted to local peculiarities of environment.

If the facts of nature now briefly sketched are clearly apprehended, there result two propositions of the highest importance. These are: (1) That whatever the amount of variability of a species, no general modification of it will occur so long as the environment remains unchanged; and (2) that when a permanent change (not a mere temporary fluctuation) of the environment occurs—whether of climate, of extension or elevation of land, of diminished food-supply, or of new competitors, or of new enemies—then, and then only, will various specific forms become modified, *so as to adapt them more completely to the new conditions of existence.* It is easy to see that all the kinds of changes above indicated are so connected that they will inevitably occur together, though in various degrees. Changes of climate or in the area or elevation of the land will cause changes in the vegetation; this will afford more or less food to various kinds of animals, and these animals will again be preyed upon by other animals. At the same time, most of these animals will need adaptation to the climate itself, whether hotter or colder, wetter or dryer; and each will also have new competitors and more or fewer enemies. There will thus be action and reaction of the most complex kind, and if the change of environment is great in amount and extends over a wide area, then, when it is completed, almost all the species that formerly existed will be found to have become more or less changed in form, structure, and habits so as to constitute new species, and in some cases new genera. This amount of change has again and again actually occurred, as shown by the extinct animals and plants preserved in the rocks of the various subdivisions of the Tertiary Period. On the other hand, we find proofs of periods of stability in the fact that certain deposits which, from their extent and thickness, must have required a long period of time for their formation, yet contain from top to bottom an almost identical series of species.

The various modifications of form, structure, or colour thus produced constitute the "specific character" of each new species, distinguishing it both from its parent species (which will usually have become extinct) and from all its near allies, and each of these characters must have been, at the time they were fixed by continued selection, *useful* to the species. This has been, and still is, denied by many naturalists, mainly because they cannot see or imagine any use in many characters; but they have never succeeded in discovering any cause, other than utility, adequate to produce new characters which shall be present in all the individuals of a species and strictly confined to them. In order to be developed through natural selection a particular variation must not only be *useful*, but must, at least occasionally, be of such importance as to lead to the saving of life, or, to use Professor Lloyd-Morgan's suggestive term, be of "survival-value." This subject has been somewhat fully discussed in Chapters V. to X. of my *Darwinism*, supplemented and enforced by the five chapters on the "Theory of Evolution" (in my *Studies*, vol. i.), and to these works I must refer my readers for fuller information.

There is, however, one other preliminary question of special importance as regards the subject of the present essay, that must be briefly discussed.

A problem of the highest importance as regards the whole theory of organic evolution, and especially in its varied applications to man's nature and advancement both individually and socially, is to determine the limits of heredity. The first great writer who put forth a detailed theory of the method of organic evolution was Lamarck, who believed that the chief cause of the modification of species was desire and effort, leading to the use of certain organs and parts to their fullest capacity; that such use strengthened and enlarged such organs (as it undoubtedly does); and that this increased development was transmitted to the offspring. In this way he thought that all the adaptations of animals to their mode of life—the strength of the lion, the speed of the antelope, the long neck of the giraffe, and all other such characters, had been acquired.

But his facts and arguments, though highly ingenious, made little impression, mainly because naturalists perceived that his theory was only applicable to a very small portion of the adaptations which needed explanation. As examples, no exercise of the will or of the muscles could produce those wonderful harmonies of colour which serve to conceal so many herbivorous creatures from their enemies and enable so many of the rapacious kinds to approach their prey. Neither could any similar action of body or mind lead to the growth of the wonderful shells of the mollusca, the bony armour of the tortoises, or the poison-glands of snakes or of stinging-insects. Still less can such causes have been effective in the production of countless adaptations in the whole structure of plants, and especially of their flowers and fruits.

But although Lamarck's theory was seen to be utterly inadequate, its fundamental assumption—that the effects of the use and disuse of organs were transmitted to the offspring—seemed so probable that it was generally accepted without any critical examination, and was even thought by Darwin to be a useful adjunct to his own theory. This was partly due to the fact that none of the early naturalists paid much attention to *variation*, which was only a source of trouble to them in their efforts to define "true species." But with the interest excited by Darwin's works the study of varieties was seen to be of the first importance, since they are the materials out of which new species are formed. It was then very soon found that variations are much more abundant, much larger in amount, and much more varied in character than was supposed, and that, together with the ever-growing proofs of the extreme rigour of natural selection, a sufficient explanation of the origin of all species was attained.

Then followed the investigations of Galton and Weismann, showing that there was no valid evidence of the transmission of the modifications of individuals due to use or disuse, or to climate, food, or other external agencies, while the elaborate researches of the latter into the earliest processes of reproduction, resulting in his illuminating theory of the "continuity of the germ-plasm," gave what was probably the death-blow to Lamarck's fundamental assumption.

During the last decade evidence has been accumulating to prove that, among the higher animals at all events, it is only the inborn characters and tendencies—whether physical or mental—that have any part in producing the varying characters of the offspring, and at the present time it may be said that almost all the chief biological thinkers and investigators hold this view, including Sir E. Ray-Lankester, Professors Lloyd-Morgan and E. B. Poulton and Sir W. T. Thiselton-Dyer. The latest and one of the acutest students of this and allied problems—Mr. G. Archdall Reid—in his remarkable work, *The Principles of Heredity*, sums up the whole case against the heredity of acquired characters with great lucidity, and shows that instead of being, as Darwin and others thought it might be, an aid to natural selection in bringing about adaptation to new conditions, it is really in the great majority of cases antagonistic to it, while in some it would actually neutralise it altogether. Those who are interested in this problem should study Mr. Reid's work, or the smaller but equally conclusive treatise by Mr. W. Platt Ball, *Are the Effects of Use and Disuse Inherited?* in which the evidence on both sides is fairly given and the answer shown to be unmistakably in the negative.

Before leaving this part of our subject it may be as well to state that, broadly speaking, every fact and argument here given applies to the mental as well as to the bodily organs, to the intellectual as well as to the physical powers of animals, including man. Galton has proved that genius, like physical qualities, is hereditary, and all other mental faculties are equally so. Great genius, like gigantic stature or enormous strength, is rare, and men in general approximate to the average in mind as well as in body. The only important difference seems to be that the mental faculties vary to a greater extent than the bodily organs. Newton and Cayley in mathematics, Shakespeare and Shelley in poetry, rise higher above the average man in intellect than do the equally rare giants in stature. This feature will be referred to later. Again, all the evidence goes to show that, though native or inborn faculties are hereditary, mental acquisitions or the results of education or experience are *not* transmitted any more than those of the bodily organs. If such acquisitions were transmitted, we should expect the younger sons of great men to possess their fathers' abilities to a greater degree than the elder sons, since they should inherit some of the ability due to the practice of the father's art or profession for a longer period; but no such constant difference has ever been detected. Equally suggestive is the fact that the children of English, French, or German parents, whose ancestors for many generations have spoken each their own tongue, do not show any exceptional power of acquiring the vocal peculiarities and intonations of their own rather than another language; and this applies in cases where the speech is most diverse and most difficult to acquire in later life, as when a European infant is reared among Chinese or Red Indians. These various considerations render it almost certain that the phenomena of heredity and variation are fundamentally the same in the mental as in the physical departments of human nature. This being the case, we must assume that Character (which is really the aggregate of the intellectual and moral faculties), in order to be progressively developed, must be acted upon by some form of natural selection. We have seen, however, that this power only acts—can only act—by the survival of individuals which possess the more *useful* developments. It follows that those special faculties which build up Character can only be preserved and increased inasmuch as they are of use to the individual or to the race, and this *utility* must be of such a nature as in times of stress or danger to be of *life-preserving value*. We must, therefore, proceed to inquire in what way and to what extent Character has been, or is being, modified or advanced.

Every evolutionist now believes that man has arisen from the lower animals by a process of modification, in the same way as any species of animal has arisen from its ancestral forms. He is the culminating point of the whole vast fabric of the organic world. If he has not so arisen, but is the product of other unknown forces guided by infinite power, then the slow development of the infinitely varied forms of nature that preceded his advent appears to be unmeaning. But if there is any purpose in the universe, if nature and man are not the chance products of primeval forces, we must conclude that the process by which man has actually been developed is the best, perhaps the only possible, mode of producing him. From this point of view everything is harmonious and intelligible. The end to be attained, required, and justified, the countless ages of preparation, of which we obtain some imperfect knowledge in the geological record. The varied forms of vegetable and animal life which filled the earth when man first appeared afforded him the means of life, not in one part only, but over the whole terrestrial surface. As he advanced in knowledge and increased in population, an ever-increasing proportion of plants and animals became of use to him, first for food, then for weapons, for clothing, for houses, for utensils; and later on for comfort and luxury, as aids to his mental development or to charm him by their beauty. The marvellous phenomena of nature, from the glittering hosts of heaven to the exquisite panorama of the seasons on the earth, awoke in him the desire for knowledge; and, as time went on, ever more and more of the secrets of nature were revealed to him, ever more and more of her powers were utilised, an ever-increasing proportion of the animal and vegetable and mineral worlds became subservient to his needs, or gratified his intellectual or æsthetic or moral faculties. Yet more, if there is a purpose in the universe, if the organic world came into existence in order that man might exist, then we must also recognise purpose in that infinite *variety* of nature, whether animate or inanimate, which has furnished such an inexhaustible supply of everything necessary for his life and happiness, and for the progressive development of his intellectual and moral nature. We can believe (and not be afraid to acknowledge our belief) that the dog and the cat, the sheep and the cow, the horse and the ass, the fowl and the pigeon, the throstle and the nightingale, the orange and the apple, the strawberry and the vine, wheat and maize, pine-tree and oak, and all the myriad luscious fruits and fragrant flowers and glorious blossoms, and infinitely varied mineral and chemical products—all alike exist as parts of the great design of human development.*

And here again we obtain further indications of purpose in the very method of organic as distinct from that of inorganic evolution. The two great distinctive features of living substance—enormous powers of increase, together with gradual but almost unlimited variability (features that are absent from the entire inorganic universe)—necessarily lead to the rapid spread of life over the whole area that is not absolutely unfitted for it, and at the same time give rise to an ever-increasing variety of forms and complexity of structure, in adaptation to the ever-changing conditions of the earth's surface. Thus the whole earth and ocean have become filled with continually varying and progressing forms of life, so that when the cosmic process culminated in man, with faculties and aspirations calculated to utilise and appreciate them, it also culminated in those highest developments of the animal and vegetable worlds which we have briefly enumerated and which certainly never existed together in equal variety and beauty at any earlier period of the earth's history.

* This subject is discussed at some length in my *World of Life* (Chapman and Hall).

But though it is admitted that man has arisen from a lower animal form, we have still to inquire whether his whole intellectual, æsthetic, and moral nature has been produced by the action of the very same laws and processes as have led to the development of animal forms and animal natures. Do variation and survival of the fittest explain man's mind as well as his body? Does he differ from the lower animals in degree only, or is there an essential difference in his mental nature?

It is clear that from brute to man there has been a great advance. This is universally admitted. But that there has been any very great advance from the earliest men of whom we have any records to ourselves, is by no means generally admitted and certainly cannot be proved.

I have myself shown that the great first step that caused man to rise above his fellow animals was that amount of mental superiority that enabled him to obtain some command over nature. After showing how each animal form could only preserve its existence in a changing universe by corresponding changes in bodily structure or in the lower mental faculties, I go on to describe what occurred in the case of man:

At length, however, there came into existence a being in whom the subtle force we term *mind*, became of more importance than his mere bodily structure. Though with a naked and unprotected body, *this* gave him clothing against the varying inclemencies of the seasons. Though unable to compete with the deer in swiftness or with the wild bull in strength, *this* gave him weapons with which to capture or overcome both. Though less capable than most other animals of living on the herbs and fruits that unaided nature supplies, this wonderful faculty taught him to govern and direct nature to his own benefit, and make her produce food for him when and where he needed. From the moment when the first skin was used for a covering, when the first rude spear was formed to assist in the chase, when fire was first used to cook his food, when the first seed was sown or shoot planted, a grand revolution was effected in nature—a revolution which in all the previous ages of the earth's history had had no parallel—for a being had arisen who was no longer necessarily subject to physical change with the changing universe, a being who was in some degree superior to nature, inasmuch as he knew how to control and regulate her action, and could keep himself in harmony with her, not by a change in body, but by an advance in mind.*

Now this passage, first published in 1864, seems to me to indicate the essential superiority of man over the lower animals, a superiority which was perhaps as great fundamentally in palæolithic or eolithic man as it is now. All that we have done since, all the triumphs of our civilisation and of our science, have arisen by slow, very slow, progressive steps, each one only a little in advance of what had been done before, and none of them perhaps so difficult, so clearly showing superiority of intellect, as those marvellous first steps which proved that a new and higher being had appeared on the earth.

Mr. Archdall Reid, in his work on Heredity already referred to (in a very suggestive chapter on "Racial Mental Differences"), adopts the views of Buckle and John Stuart Mill, that by far the larger part of racial or national differences of character are *not* inherent, but are the product of the diverse and highly complex environments of each. This would include, of course, their past history, their religion, their education, their form of government, and the various habits and customs, language, legends, and superstitions that have come down to them from a forgotten past. In comparing a savage with a civilised race, we must always remember that the amount of acquired and applied knowledge which we possess is no criterion of mental superiority on our side, or of inferiority on his. The average Zulu or Fijian may be very little lower mentally than the average Englishman; and it is, I think, quite certain that the average Briton, Saxon, Dane, and Norseman of a thousand years ago—the ancestral stocks of the present English race—were mentally our equals. For what power has been since at work to improve them? There has certainly been no special survival of the more intellectual and moral, but rather the reverse. As Galton points out, the celibacy of the Roman Church and the seclusion of thousands of the more refined women in abbeys and nunneries tended to brutalise the race.

To this we must add the destruction of thousands of psychics, many of them students and inventors, during the witchcraft mania, and the repression of thought and intellect by the Inquisition; and when we consider further that the effects of education and the arts are not hereditary, we shall be forced to the conclusion that we are to-day, in all probability, mentally and morally inferior to our semi-barbaric ancestors!

* *Natural Selection and Tropical Nature*, 181.

Looking back at the course of our history from the Saxon invasion to the end of the nineteenth century, what single cause can we allege for an advance in intellect and in moral nature? What selective agency of "survival value" has ever been at work to preserve the wise and good and to eliminate the stupid and the bad? And it must have certainly been a very powerful agency, acting in a very systematic manner, even to neutralise the effect of the powerful deteriorating agencies above referred to.

When we remember that the Romans and the Greeks looked down on *all* our ancestors as we look down on Kaffirs and Red Indians, we must not too hastily conclude that, because people are in the savage or barbarian state as regards knowledge and material civilisation, they are necessarily inferior intellectually or morally. I am inclined to believe that an unbiassed examination of the question would lead us to the conclusion which, as I understand him, is favoured by Mr. Archdall Reid, and that there is no good evidence of any considerable improvement in man's average intellectual and moral status during the whole period of human history, nor any difference at all in that status corresponding with differences in material civilisation between civilised and savage races today. What differences actually exist are sufficiently accounted for by various selective agencies known to have been at work; while there is good reason to believe that some of the lowest savages to-day (perhaps all of them) are the deteriorated remnants of more civilised peoples.

If we turn to the facts actually known to us about early man, historic and prehistoric, they certainly point in the same direction. Whence came the wonderful outgrowth of art manifested by the Germans and Celts in their Gothic architecture, admirable alike in structure, in design, and in ornament, and which we, however much we pride ourselves on our science, cannot approach in either originality or beauty? Going further back, the Roman architects, sculptors, poets, and literary men were fully our equals. Still earlier, the Greeks were our equals, if not superior in art, in literature, and in philosophy. The Aryans of Northern India were equally advanced, and their wonderful epic—the Maha-Bharata—introduces us to a people who were probably both in intellect and in morality no whit inferior to ourselves. Further back still, in ancient Egypt, we find in the Great Pyramid a structure which is the oldest in the world, and in many respects the most remarkable. In its geometrical proportions, its orientation, and its marvellous accuracy of construction, it is in itself the record of a people who had already attained to a degree of high intellectual achievement. It was one of the most gigantic astronomical observatories ever erected by man, and it shows such astronomical and geometrical knowledge, such precision of structure, and such mechanical skill, as to imply long ages of previous civilisation, and an amount of intellect and love of knowledge fully equal to that of the greatest mathematicians, astronomers, and engineers of our day.

And if from the purely intellectual we turn to the domain of conduct and of ethical standards, we encounter facts which also lead us to the same conclusion. If we compare the two greatest ethical teachers of our age with their earliest prototypes whose works have been preserved, it is impossible to maintain that there has been any real advance in their respective characters. Tolstoy can hardly be ranked as higher than Buddha, or Ruskin than Confucius; and as we cannot suppose the amount of variation of human faculty about a mean value to be very different now from what it was at that remote era, we must conclude that equality in the highest implies equality in the mean, and that human nature on the whole has not advanced in intellect or in moral standards during the last three thousand years, while the records of Egypt in both departments—those of science and of ethics—enable us to extend the same conclusion to a period some thousands of years earlier.

In reply to this argument, it will be urged that the period from these early civilisations to our own day is only a fragment of man's whole history, and that in the remains of neolithic, palæolithic, and eolithic man we have certain proofs that his earliest condition was that of a low and brutal savage. But this is pure assumption, because the evidence at our command does not bear upon the question at issue. Material civilisation and inherent character are things which have no necessary connection. There is no inconsistency, no necessary contradiction, in the supposition that the men of the early stone age were our equals intellectually and morally. As Mr. Archdall Reid well argues, if a potential Newton or Darwin were occasionally born among savages, how could his faculties manifest themselves in that forbidding environment? With an imperfect language and limited notation, and having to maintain a constant struggle for existence against the forces of nature, and in combination with his fellows against wild beasts and human enemies, either of them might have made some one step in advance— might have invented some new weapon or constructed some improved trap. He must necessarily work on the lines of his fellows and with the materials to his hand. Perhaps in the rude drawings of animals on stone or tusk we have the work of a potential Landseer; while the equal of our Watt or Kelvin might have initiated the polished stone axe or invented the bone needle. That a people without metals and without written language, who could therefore leave few imperishable remains, may yet possess an intellect and moral character fully equal (some observers think superior) to our own, is demonstrated in the case of the Samoans, and some other tribes of the Pacific. It is clear, therefore, that a low state of material civilisation is no indication whatever of inferiority of character.

But although every indication of history and of existing races of man negatives the idea of any general *advance* of character—a conclusion which is supported by the entire absence of any selective agency of "survival value," which could alone have led to such advance on the principles of organic evolution—yet there are undoubtedly *differences* of national character which it is not easy to account for. That, on the whole, the Celtic races are more idealistic, more joyous, and more excitable than the Germanic or Sclavonian, while a similar difference exists between the peoples of Southern and Northern Europe, seems to be generally admitted. Buckle, as has been already noted, explained the difference by the influence of the diverse environments, and Mr. Archdall Reid favours the same view, but there are many difficulties connected with such a theory. No doubt the best-known Celtic races—the Bretons, the Welsh, the Cornish, and the Highlanders—have been long the inhabitants of mountainous districts to which they have been driven by the invasions of more warlike peoples; but, unless some form of selection comes into play, it is difficult to imagine why this should have changed the character of people who had presumably lived at some earlier period in less awe-inspiring lands.

But the great argument against this explanation is to be found, I think, in the diverse characters of two of the principal divisions of mankind—the Mongoloid and the Negroid. Here we see that great changes of natural environment have produced no corresponding modification of character, and *vice versa*. Every reader of my *Malay Archipelago* will, I think, remember my description of the Ke Islanders (typical Papuans and Negroids) and my comparison of their behaviour with that of the Malays (equally typical Mongoloids), with whose character I was so well acquainted. Now, the fundamental features of the *characters* of these two great divisions of mankind maintain themselves wherever they are found, in every variety of aspect and of climate, extending over three-fourths of the globe. The Red Indians of America (true Mongoloids for the most part) have the same impassive, unexcitable character in the frigid, the tropical, and the temperate zones; whether they inhabit the forests or the plains, the great river valleys or the lofty plateau; and the same may be said of the Old World branch from the Japanese and Chinese to the Kalmucks and Malays; and throughout these vast diversities of natural environment it cannot be said that any minor diversities of character can be positively traced to local influences. In the case of the Malays and Papuans, we have the two races existing under almost identical circumstances in the vast equatorial forests extending from Sumatra to the Solomon Islands, often living in a very similar manner and in an almost identical stage of barbarism; yet it is in this very region that their distinctive mental characteristics are found to be at a maximum. For such a mental divergence as these two races present, I cannot myself see any possibility of an explanation through any selective agency of "survival value"; while the influence of environment is equally untenable, besides being in direct opposition to the now well-established principle of the non-heredity of acquired peculiarities.

In this undoubted difference of racial character, and perhaps to an even greater extent in that of national character, the mental divergence seems to exceed the physical. The former better corresponds to the amount of mental difference between different species, genera, or even families of animals than to those presented by mere varieties of a single species; and in this way we have an indication of a want of parallelism or of direct relationship of the mental and the physical characteristics of mankind, which may, perhaps, offer us a clue to this most complex and important problem. Among individuals we see the same phenomenon, though we have no means of accurately estimating it. The amount of divergence in the physical features of healthy and equally well-nourished and well-trained individuals in the same country is not very great. In stature, strength, speed of running, and acuteness of the senses, the divergence from the mean is rarely more than as two to three, and in the most extreme cases does not exceed two to one. But in the mental faculties, or in any special faculty, the divergence would be usually estimated at a much higher figure. There are thousands of mathematicians among us to-day whose capacities would certainly be estimated at five or six times greater than that of other thousands who can never understand comparatively simple arithmetical or geometrical problems, while the extreme cases of the highest mathematical genius and the lowest degree of arithmetical stupidity would be estimated as at least some such proportion as 100 to 1, if not much higher; and in every other department of human faculty—music, poetry, or eloquence—there is perhaps a nearly equal amount of divergence.

We may, I think, explain this circumstance by the consideration that, while the physical characteristics must have been rigidly selected during the earlier period of man's existence on the earth, through his constant struggle with the lower animals and against the forces of nature, and later on almost equally so in war or competition with other tribes or races, his higher mental faculties were seldom or never called into action, being of no direct use to him in the struggle for existence. While the former, therefore, became fixed within definite limits, the latter were free to vary in amount through the agency of some unknown laws or inherent capacities. The extraordinary thing is, that these higher faculties did not become atrophied by disuse as would physical characters under similar conditions; instead of which they appear to have persisted undiminished in power throughout all human history, ready under favourable conditions to blaze out in a Homer or a Socrates, a Pyramid designer or a Buddha, an Archimedes or a Shakespeare.

Some of the greatest upholders of the theory of natural selection admit that these higher faculties cannot have been developed through its agency. In an elaborate essay on *The Musical Sense in Man and Animals*, Weismann comes to the conclusion that the musical sense "is simply a by-product or accessory of the auditory organ," and that it is "a merely incidental production, and thus, in a certain sense, an unintended one." In another work (his lecture on *Heredity*) he arrives at a similar conclusion with regard to all the higher activities of the mind in the following statement: "In my opinion, talents do not appear to depend upon the improvement of any special mental quality by continual practice, but they are the expression, and to a certain extent the by-product, of the human mind, which is so highly developed in all directions." Huxley arrived at a somewhat similar conclusion, being reported by Mr. Wilfred Ward as saying: "One thing which weighs with me against pessimism, and tells for a benevolent author of the universe, is, my enjoyment of scenery and of music. I do not see how they can have helped in the struggle for existence. They are gratuitous gifts."

But though there has, apparently, been no continuous advance in the higher intellectual and moral nature of man for want of any selective agency leading to such a result, this has not been the case with that portion of his faculties which he possesses in common with the lower animals. The family affections, and the social instincts, were essential to the safety of the clan or the tribe; courage and perseverance, cautiousness and decision, were valuable in hunting and in war; the inventive and constructive faculties were of value in the making of weapons and snares, clothing and houses, while foresight, and the love of animals, might lead to the simpler forms of agricultural industry. These, however, could hardly have arisen till after the invention of weapons and of tools, as well as the discovery of the use of fire, and it is by no means easy to see how natural selection alone, which can only produce modifications in accordance with an animal's needs, never beyond them, could have led to that mental superiority which at once placed man so far above all other animals, and have endowed him with such capacities for advancement as he actually possessed.

From the sketch which has now been given of the actual powers of the human mind, and of the various influences which may conceivably have modified it, we have been led to some very startling conclusions. We see, first, that the general idea that our enormous advances in science and command over nature serve as demonstrations of our mental superiority to the men of earlier ages, is totally unfounded. The evidence of history and of the earliest monuments alike go to indicate that our intellectual and moral nature has not advanced in any perceptible degree. In the second place, we find that the supposed great mental inferiority of savages is equally unfounded. The more they are sympathetically studied, the more they are found to resemble ourselves in their inherent intellectual powers. Even the so-long-despised Australians—almost the lowest in material progress—yet show by their complex language, their elaborate social regulations, and often by an innate nobility of character, indications of a very similar inner nature to our own. If they possess fewer philosophers and moralists, they are also free from so large a proportion of unbalanced minds—idiots and lunatics—as we possess. On the other hand, we find in the higher Pacific types men who, though savages as regards material progress, are yet generally admitted to be—physically, intellectually, and morally—our equals, if not our superiors. These we are rapidly exterminating through the effect of *our* boasted civilisation!

Thirdly, we have no proof whatever that even the men of the stone age were mentally or morally inferior to ourselves. The case of the Pacific Islanders shows that simple arts and constructions with the absence of written language affords no proof of inferiority; while the undoubted absence of any selective power of "survival value" adequate to the evolution of the higher intellectual, æsthetic, and moral faculties—which we find so fully developed in Ancient India, Egypt, and Greece—indicates that the very earliest men of whose existence we have any certain knowledge must also have possessed these faculties. If they did not possess them there must have been progressive mental progress independent of selection and without any intelligible cause.

One other characteristic of man which supports this conclusion is, as already shown, the extreme variability of his whole mental and moral nature, a variability much greater than that present in his body; and this again indicates that there has been no selective agency adequate to limit its range or guide it in any special direction.

The preceding considerations lead us to conclude that the higher mental or spiritual nature of man is not the mere animal nature advanced through survival of the fittest. All the greatest writers and thinkers on the subject now admit this. In the last chapter of my *Darwinism* I have shown that some of the bodily characteristics of man are similarly inexplicable as the result of the same selective process. Darwin himself declared that the law of natural selection was, in his opinion, the greatest but not the exclusive means of modification.

To me it appears that, just as gravitation rules the whole material universe, so natural selection rules, and has ruled, the whole organic world. But in the countless modifications of matter, other quite distinct forces control or antagonise gravitation. Molecular and chemical forces, within their sphere of action, are far more powerful, and entirely neutralise the effect of the more far-reaching agency. Electricity and other ethereal forces are still more powerful; and, as seen at work in cometary emanations, oppose and overcome the gravitative force of the sun.

In like manner we see in the organic world a new and higher series of powers at work. First, in the life-force that renders possible the whole marvellous structure, growth, and products of the vegetable kingdom; next, in the higher life of consciousness and purposive action, as manifested in the lower animals; and, lastly, in the still higher spiritual nature—a little lower than that of the angels—with which man is endowed.

There is no evidence whatever that any of the animal forms below man possess the germs of this higher nature, however intelligent and teachable many of them undoubtedly are. Had they possessed it, some of them would have given indications of it. Such very diverse animals as the cat and the dog, the horse and the elephant, the monkey and the chimpanzee, exhibit a nearly equal amount of animal intelligence, but none of them can be said to be decidedly superior to the rest, none show any clear signs of the possession of even the rudiments of those faculties which raise man so infinitely above them all, or even of those much lower yet still essential powers, which enabled man, as soon as he became man, to develop language, to utilise fire, to make tools and weapons, to sow seeds, and to become shepherds and herdsmen.

But from the epoch when man first attained to his specially human powers, he not only at once assumed command over the earth and all its forms of life, but commenced that development of latent faculties of which we find such striking evidences throughout history. The whole universe, in all its myriad forms, in all its intricacies of structure and motion, in all its marvellous beauty and inexhaustible utilities, in all its complex and mysterious laws and forces, became to him a vast school-room, furnishing the materials needed for the development of all his hitherto unused faculties and for the gradual elevation of his intellectual and moral nature. But the possibilities of such development must have pre-existed; the germs must have been present; every faculty must have been latent, or no amount of marvel and mystery could have called them forth.

How this higher nature originated in man, we may never know; but all the evidence points in the direction of some spiritual influx analogous to that which first initiated the organised life of the plant; then the consciousness and intelligence of the animal; and, lastly, reason, the sense of beauty, the love of justice, the passion for truth, the aspiration towards a higher life which everywhere, though in varying degrees, characterise the Human Race.

Although, as we have seen, there has been no general advance of Character during the whole period of which we can obtain any definite information, due to the absence of any great or constant selective agency, there is every reason to believe that it will be so improved in the not distant future. The heights to which it has attained in a few rare examples in all ages, taken in connection with the enormous range of variation it presents at the present time, show us that ample materials exist for raising its present average almost indefinitely. This can be effected by two distinct influences, which can and must always work together—education and selection by marriage. As yet we have no true and effective education. The very first essential in the teacher—true love of and sympathy with the children—is not made one of the conditions of entering that great profession. Till this is made the *primary* qualification (as it was by Robert Owen at his schools in New Lanark) no real improvement in social and moral character can be effected. Mere intellectual instruction—which is all now given—is *not* a complete education, is really the least important half of it.

The other and more permanently effective agency, selection through marriage, will come into operation only when a greatly improved social system renders all our women economically and socially free to choose; while a rational and complete education will have taught them the importance of their choice both to themselves and to humanity. This subject I have treated in my *Studies* (see the chapter on "Human Selection"). It will act through the agency of well-known facts and principles of human nature, leading to a continuous reduction of the lower types in each successive generation, and it is the only mode yet suggested which will automatically and naturally effect this.

When we consider the enormous importance of such a continuous improvement in the average character, and that our widespread and costly religious and educational agencies have, so far, made not the slightest advance towards it, we shall, perhaps, realise, before it is too late, that we have begun at the wrong end. Improvement of social conditions must precede improvement of Character; and only when we have so reorganised society as to abolish the cruel and debasing struggle for existence and for wealth that now prevails, shall we be enabled to liberate those beneficent natural forces which alone can elevate Character.

The great lesson taught us by this brief exposition of the phenomena of Character in relation to the known laws of organic evolution is this: that our imperfect human nature, with its almost infinite possibilities of good and evil, can only make a systematic advance through the thoroughly sympathetic and ethical training of every child from infancy upwards, combined with that perfect freedom of choice in marriage which will only be possible when all are economically equal, and no question of social rank or material advantage can have the slightest influence in determining that choice.

When our workers, our thinkers, our legislators can be persuaded to accept these fundamental truths, and make them the twin guiding stars of their aspirations and their efforts, the onward march towards true civilisation will have begun, and for the first time in the history of mankind, his Character—his very Human Nature itself—will be improved by the slow but certain action of a pure and beautiful form of selection—a selection which will act, not through struggle and death, but through brotherhood and love.

∽ ✌

Impressions of the United States

Having now left North America, I may say just a few words of my general impressions as to the country and the people. In my journal I find this note: "During more than ten months in America, taking every opportunity of exploring woods and forests, plains and mountains, deserts and gardens, between the Atlantic and Pacific coasts, and extending over ten degrees of latitude, I never once saw either a humming-bird or a rattlesnake, or even any living snake of any kind. In many places I was told that humming-birds were usually common in their gardens, but they hadn't seen any this year! This was my luck. And as to the rattlesnakes, I was always on the lookout in likely places, and there are plenty still, but they are local. I was told of a considerable tract of land not far from Niagara which is so infested with them that it is absolutely useless. The reason is that it is very rocky, with so many large masses lying about overgrown with shrubs and briars as to afford them unlimited hiding-places, and the labour of thoroughly clearing it would be more costly than the land would be worth.

The general impression left upon my mind as to the country itself is the almost total absence of that simple rural beauty which has resulted, in our own country and in some other parts of Europe, from the very gradual occupation of the land as it was required to supply food for the inhabitants,

My Life, 2:190–99.

together with our mild winters allowing of continuous cultivation, and the use in building of local materials adapted to the purposes required by hand-work, instead of those fashioned by machinery. This slow development of agriculture and of settlement has produced almost every feature which renders our country picturesque or beautiful: the narrow winding lanes, following the contours of the ground; the ever-varying size of the enclosures, and their naturally curved boundaries; the ditch and the bank and the surmounting hedgerow, with its rows of elm, ash, or oak, giving variety and sylvan beauty to the surroundings of almost every village or hamlet, most of which go back to Saxon times; the farms or cottages built of brick or stone, or clay, or of rude but strong oak framework filled in with clay or lath and roughcast, and with thatch or tiled roofs, varying according to the natural conditions, and in all showing the slight curves and irregularities due to the materials used and the hand of the worker—the whole, worn and coloured by age and surrounded by nature's grandest adornment of self-sown trees in hedgerow or pasture, combine together to produce that charming and indescribable effect we term picturesque. And when we add to these the numerous footpaths which enable us to escape the dust of high roads and to enjoy the glory of wild flowers which the innumerable hedgerows and moist ditches have preserved for us, the breezy downs, the gorse-clad commons and the heath-clad moors still unenclosed, we are, in some favoured districts at least, still able thoroughly to enjoy all the varied aspects of beauty which our country affords us, but which are, alas! under the combined influences of capitalism and landlordism, fast disappearing.

But in America, except in a few parts of the north-eastern States, none of these favourable conditions have prevailed. Over by far the greater part of the country there has been no natural development of lanes and tracks and roads as they were needed for communication between villages and towns that had grown up in places best adapted for early settlement; but the whole country has been marked out into sections and quarter-sections (of a mile, and a quarter of a mile square), with a right of way of a certain width along each section-line to give access to every quarter-section of one hundred and sixty acres, to one of which, under the homestead law, every citizen had, or was supposed to have, a right of cultivation and possession. Hence, in all the newer States there are no roads or paths whatever beyond the limits of the townships, and the only lines of communication for foot or horsemen or vehicles of any kind are along these rectangular section-lines, often going up and down hill, over bog or stream, and almost always compelling the traveler to go a much greater distance than the form of the surface rendered necessary.

Then again, owing to the necessity for rapidly and securely fencing in these quarter-sections, and to the fact that the greater part of the States first settled were largely forest-clad, it became the custom to build rough, strong fences of split-trees, which utilized the timber as it was cut and involved no expenditure of cash by the settler. Again, to avoid the labour of putting posts in the ground the fence was at first usually built of rails or logs laid zigzag on each other to the height required, so as to be self-supporting, the upper pairs only being fastened together by a spike through them, the waste of material in such a fence being compensated by the reduction of the labour, since the timber itself was often looked upon as a nuisance to be got rid of before cultivation was possible. And yet again, this fact of timber being in the way of cultivation and of no use till cut down, led to the very general clearing away of all the trees from about the house, so that it is a comparatively rare thing, except in the eastern towns and villages, to find any old trees that have been left standing for shade or for beauty.

For these and for similar causes acting through the greater part of North America, there results a monotonous and unnatural ruggedness, a want of harmony between man and nature, the absence of all those softening effects of human labour and human occupation carried on for generation after generation in the same simple way, and in its slow and gradual utilization of natural forces allowing the renovating agency of vegetable and animal life to conceal all harshness of colour or form, and clothe the whole landscape in a garment of perennial beauty.

Over the larger part of America everything is raw and bare and ugly, with the same kind of ugliness with which we also are defacing our land and destroying its rural beauty. The ugliness of new rows of cottages built to let to the poor, the ugliness of the mean streets of our towns, the ugliness of our "black countries" and our polluted streams. Both countries are creating ugliness, both are destroying beauty; but in America it is done on a larger scale and with a more hideous monotony. The more refined among the Americans see themselves as clearly as we see it. One of them has said, "A whole huge continent has been so touched by human hands that, over a large part of its surface it has been reduced to a state of unkempt, sordid ugliness; and it can be brought back into a state of beauty only by further touches of the same hands more intelligently applied."*

Turning now from the land to the people, what can we say of our American cousins as a race and as a nation? The great thing to keep in mind is, that they are, largely and primarily, of the same blood and of the same nature as ourselves, with characters and habits formed in part by the evil traditions inherited from us, in part by the influence of the new environment to which they have been exposed. Just as we owe our good and bad qualities to the intermixture and struggle of somewhat dissimilar peoples, so do they. Briton and Roman, Saxon and Dane, Norsemen and Norman-French, Scotch and Irish Celts—all have intermingled in various portions, and helped to create that energetic amalgam known the world over as Englishmen. So North America has been largely settled by the English, partly by Dutch, French, and Spanish, whose territories were soon absorbed by conquest or purchase; while, during the last century, a continuous stream of immigrants—Germans, Irish, Highland and Lowland Scotch, Scandinavians, Italians, Russians—has flowed in, and is slowly but surely becoming amalgamated into one great Anglo-American people.

Most of the evil influences under which the United States have grown to their present condition of leaders in civilization, and a great power among the nations of the world, they received from us. We gave them the example of religious intolerance and priestly rule, which they have now happily thrown off more completely than we have done. We gave them slavery, both white and black—a curse from the effects of which they still suffer, and out of which a wholly satisfactory escape seems as remote as ever. But even more insidious and more widespread in its evil results than both of these, we gave them our bad and iniquitious feudal land system; first by enormous grants from the Crown to individuals or to companies, but also—what has produced even worse effects—the ingrained belief that *land*—the first essential of life, the source of all things necessary or useful to mankind, by labour upon which all wealth arises—may yet, justly and equitably, be owned by individuals, be monopolized by capitalists or by companies, leaving the great bulk of the people as absolutely dependent on these monopolists for permission to work and to live as ever were the negro slaves of the South before emancipation.

* *The Century*, June 1887.

The result of acting upon this false conception is that the Government has already parted with the whole of the accessible and cultivable land, and though large areas still remain for any citizen who will settle upon it by the mere payment of very moderate fees, this privilege is absolutely worthless to those who most want it—the very poor. And throughout the western half of the Union one sees everywhere the strange anomaly of building lots in small remote towns, surrounded by thousands of uncultivated acres (and perhaps ten years before sold for eight or ten shillings an acre), now selling at the rate of from £1,000 to £20,000 an acre! It is not an uncommon thing for town lots in new places to double their value in a month, while a four-fold increase in a year is quite common. Hence land speculation has become a vast organized business over all the Western States, and is considered to be a proper and natural mode of getting rich. It is what the Stock Exchange is to the great cities. And this wealth, thus gained by individuals, initiates that process which culminates in railroad and mining kings, in oil and beef trusts, and in the thousand millionaires and multi-millionaires whose vast accumulated incomes are, every penny of them, paid by the toiling workers, including the five million of farmers whose lives of constant toil only result for the most part in a bare livelihood, while the railroad magnates and corn speculators absorb the larger portion of the produce of their labour.

What a terrible object-lesson is this as to the fundamental wrong in modern societies which leads to such a result! Here is a country more than twenty-five times the area of the British Islands, with a vast extent of fertile soil, grand navigable waterways, enormous forests, a superabounding wealth of minerals—everything necessary for the support of a population twenty-five times that of ours—about fifteen hundred millions—which has yet, in little more than a century, destroyed nearly all its forests, is rapidly exhausting its marvellous stores of natural oil and gas, as well as those of the precious metals; and as the result of all this reckless exploiting of nature's accumulated treasures has brought about over-crowded cities reeking with disease and vice, and a population which, though only one-half greater than our own, exhibits all the pitiable phenomena of women and children working long hours in factories and workshops, garrets and cellars, for a wage which will not give them the essentials of mere healthy animal existence; while about the same proportion of its workers, as with us, endure lives of excessive labour for a bare livelihood, or constitute that cry-ing disgrace of modern civilization—willing men seeking in vain for honest work, and forming a great army of the unemployed.

What a demonstration is this of the utter folly and stupidity of those blind leaders of the blind who impute all the evils of *our* social system, all *our* poverty and starvation, to over-population! Ireland, with half the population of fifty years ago, is still poor to the verge of famine, and is therefore still overpeopled. And for England and Scotland as well, the cry is still, "Emigrate! emigrate! We are over peopled!" But what of America, with twenty-five times as much land as we have, with even greater natural resources, and with a population even more ingenious, more energetic, and more hardworking than ours? Are they over-populated with only twenty people to the square mile? There is only one rational solution of this terrible problem. The system that allows the land and the minerals, the means of communication, and all other public services, to be monopolized for the aggrandisement of the few—for the creation of millionaires—necessarily leads to the poverty, the degradation, the misery of the many.

There never has been, in the whole history of the human race, a people with such grand opportunities for establishing a society and a nation in which the products of the general labour should be so distributed as to produce general well-being. It wanted but a recognition of the fundamental principle of "equality of opportunity," tacitly implied in the Declaration of Independence. It wanted but such social arrangements as would ensure to every child the best nurture, the best training of all its faculties, and the fullest opportunity for utilizing those faculties for its own happiness and for the common benefit. Not only equality before the law, but equality of opportunity, is the great fundamental principle of social justice. This is the teaching of Herbert Spencer, but he did not carry it out to its logical consequence—the inequity, and therefore the social immorality of wealth-inheritance. To secure equality of opportunity there must be no inequality of initial wealth. To allow one child to be born a millionaire and another a pauper is a crime against humanity, and, for those who believe in a deity, a crime against God!*

It is universally admitted that very great individual wealth, whether inherited or acquired, is beneficial neither to the individual nor to society. In the former case it is injurious, and often morally ruinous to the possessor;

* I have discussed this subject in my "Studies," vol. II; chap. xxviii.

in the latter it confers little or no happiness to the acquirer of it, and is a positive injury to his heirs and a danger to the State. Yet its fascinations are so great that, under conditions of society in which the yawning gulf of poverty is ever open beside us, the amassing of wealth at first seems a duty, then becomes a habit, and, ultimately, the gambler's excitement without which he cannot live. The struggle for wealth and power is always exciting, and to many is irresistible. But it is essentially a degrading struggle, because the few only can succeed while the many must fail; and where *all* are doing their best in their several ways, with their special capacities and their unequal opportunities, the result is very much of a lottery, and there is usually no real merit, no specially high intellectual or moral quality in those that succeed.

It is the misfortune of the Americans that they had such a vast continent to occupy. Had it ended at the line of the Mississippi, agricultural development might have gone on more slowly and naturally, from east to west, as increase of population required. So again, if they had had another century for development before railways were invented, expansion would necessarily have gone on more slowly, the need for good roads would have shown that the rectangular system of dividing up new lands was a mistake, and some of that charm of rural scenery which we possess would probably have arisen.

But with the conditions that actually existed we can hardly wonder at the result. A nation formed by emigrants from several of the most energetic and intellectual nations of the old world, for the most part driven from their homes by religious persecution or political oppression, including from the very first all ranks and conditions of life—farmers and mechanics, traders and manufacturers, students and teachers, rich and poor—the very circumstances which drove them to emigrate led to a natural selection of the *most* energetic, the *most* independent, in many respects the *best* of their several nations. Such a people, further tried and hardened by two centuries of struggle against the forces of nature and a savage population, and finally by a war of emancipation from the tyranny of the mother country, would almost necessarily develop both the virtues, the prejudices, and even the vices of the parent stock in an exceptionally high degree. Hence, when the march of invention and of science (to which they contributed their share) gave them the steamship and the railroad; when California gave them gold and Nevada silver, with the prospect of wealth to the lucky beyond the dreams of avarice; when the great prairies of the West gave them illimitable acres of marvellously fertile soil—it is not surprising that these conditions with such a people should have resulted in that mad race for wealth in which they have beaten the record, and have produced a greater number of multi-millionaires than all the rest of the world combined, with the disastrous results already briefly indicated.

But this is only one side of the American character. Everywhere there are indications of a deep love of nature, a devotion to science and to literature fully proportionate to that of the older countries; while in inventiveness and in the applications of science to human needs they have long been in the first rank. But what is more important, there is also rapidly developing among them a full recognition of the failings of our common social system, and a determination to remedy it. As in Germany, in France, and in England, the socialists are becoming a power in America. They already influence public opinion, and will soon influence the legislatures. The glaring fact is now being widely recognized that with them, as with all the old nations of Europe, an increase in wealth and in command over the powers of nature such as the world has never before seen, has *not* added to the true well-being of any part of society. It is also indisputable that, as regards the enormous masses of the labouring and industrial population, it has greatly increased the numbers of those whose lives are "below the margin of poverty," while, as John Stuart Mill declared many years ago, it has not reduced the labour of any human being.

An American (Mr. Bellamy) gave us the books that first opened the eyes of great numbers of educated readers to the practicability, the simplicity, and the beauty of socialism. It is to America that the world looks to lead the way towards a just and peaceful modification of the social organism, based upon a recognition of the principle of Equality of Opportunity, and by means of the Organization of Labour of all for the Equal Good of all.

~∾ ∿~

Remembrances of Alfred Russel Wallace by His Children William G. and Violet Wallace

In our father's youth and prime he was 6 ft. 1 in. in height, with square though not very broad shoulders. At the time to which our first clear rec-

Wallace: Letters and Reminiscences by James Marchant, 1916, 349–56.

ollections go back he had already acquired a slight stoop due to long hours spent at his desk, and this became more pronounced with advancing age; but he was always tall, spare and very active, and walked with a long easy swinging stride which he retained to the end of his life.

As a boy he does not appear to have been very athletic or muscularly strong, and his shortsightedness probably prevented him from taking part in many of the pastimes of his school-fellows. He was never a good swimmer, and he used to say that his long legs pulled him down. He was, however, always a good walker and, until quite late in life, capable of taking long country walks, of which he was very fond.

He was very quick and active in his movements at times, and even when 90 years of age would get up on a chair or sofa to reach a book from a high shelf, and move about his study with rapid strides to find some paper to which he wished to refer.

When out of doors he usually carried an umbrella, and in the garden a stick, upon which he leaned rather heavily in his later years. His hair became white rather early in life, but it remained thick and fine to the last, a fact which he attributed to always wearing soft hats. He had full beard and whiskers, which were also white. His eyes were blue and his complexion rather pale. He habitually wore spectacles, and to us he never looked quite natural without them. Towards the end of his life his eyes were subject to inflammation, and the glasses were blue. His hands, though large, were not clumsy, and were capable of very delicate manipulations, as is shown by his skill in handling and preserving insects and bird-skins, and also in sketching, where delicacy of touch was essential. His handwriting is another example of that; it remained clear and even to the end, in spite of the fact that he wrote all his books, articles, and letters with his own hand until the last few years, when he occasionally had assistance with his correspondence; but his last two books, "Social Environment" and "The Revolt of Democracy," written when he was 90 years of age, were penned by himself, and the MSS. are perfectly legible and regular.

He was very domestic, and loved his home. His interest extended to the culinary art, and he was fond of telling us how certain things should be cooked. This became quite a joke among us. He was very independent, and it never seemed to occur to him to ask to have anything done for him if he could do it himself—and he could do many things, such as sewing on buttons and tapes and packing up parcels, with great neatness. When unpacking parcels he never cut the string if it could be untied, and he would fold it up before removing the paper, which in its turn was also neatly folded.

His clothes were always loose and easy-fitting, and generally of some quiet-coloured cloth or tweed. Out of doors he wore a soft black felt hat rather taller than the clerical pattern, and a black overcoat unless the weather was very warm. He wore no ornaments of any kind, and even the silver watch-chain was worn so as to be invisible. He wore low collars with turned-down points and a narrow black tie, which was, however, concealed by his beard. He was not very particular about his personal appearance, except that he always kept his hair and beard well brushed and trimmed.

In our early days at Grays we children were allowed to run in and out of his study; but if he was busy writing at the moment we would look at a book until he could give us his attention. His brother in California sent him a live specimen of the lizard called the "horned toad," and this creature was kept in the study, where it was allowed to roam about, its favourite place being on the hearth.

About this time he read "Alice through the Looking-glass," which pleased him greatly; he was never tired of quoting from it and using some of Lewis Carroll's quaint words till it became one of our classics.

Some of our earliest recollections are of the long and interesting walks we took with our father and mother. He never failed to point out anything of interest and tell us what he knew about it, and would answer our numerous questions if possible, or put us off with some joking reference to Boojums or Jabberwocks. We looked upon him as an infallible source of information, not only in our childhood, but to a large extent all his life. When exploring the country he scorned "trespass boards." He read them "Trespassers will be persecuted," and then ignored them, much to our childish trepidation. If he was met by indignant gamekeepers or owners, they were often too much awed by his dignified and commanding appearance to offer any objection to his going where he wished. He was fond of calling our attention to insects and to other objects of natural history, and giving us interesting lessons about them. He delighted in natural scenery, especially distant views, and our walks and excursions were generally taken with some object, such as finding a bee orchis or a rare plant, or exploring a new part of the country, or finding a waterfall.

In 1876 we went to live at Dorking, but stayed there only a year or two.

An instance of his love of mystifying us children may be given. It must have been shortly after our arrival at Dorking that one day, having been out to explore the neighbourhood, he returned about tea-time and said, "Where do you think I have been? To Glory!" Of course we were very properly excited, and plied him with questions, but we got nothing more out of him then. Later on we were taken to see the wonderful place called "Glory Wood"; and it had surely gained in glory by such preparation.

Sometimes it would happen that a scene or object would recall an incident in his tropical wanderings and he would tell us of the sights he had seen. At the time he was greatly interested in botany, in which he was encouraged by our mother, who was an ardent lover of flowers; and to the end of his life he exhibited almost boyish delight when he discovered a rare plant. Many walks and excursions were taken for the purpose of seeing some uncommon plant growing in its natural habitat. When he had found the object of his search we were all called to see it. During his walks and holidays he made constant use of the one-inch Ordnance Maps, which he obtained for each district he visited, planning out our excursions on the map before starting. He had a gift for finding the most beautiful walks by means of it.

In 1878 we moved to Croydon, where we lived about four years [Figure 23]. It was at this time that he hoped to get the post of Superintendent of Epping Forest. We still remember all the delights we children were promised if we went to live there. We had a day's excursion to see the Forest, he with his map finding out the roads and stopping every now and then to admire a fresh view or to explain what he would do if the opportunity were given him. It was a very hot day, and we became so thirsty that when we reached a stream, to our great joy and delight he took out of his pocket, not the old leather drinking-cup he usually carried, but a long piece of black india-rubber tubing. We can see him now, quite as pleased as we were with this brilliant idea, letting it down into the stream and then offering us a drink! No water ever tasted so nice! Our mother used to be a little anxious as to the quality of the water, but he always put aside such objections by saying *running* water was quite safe, and somehow we never came to any harm through it. The same happy luck attended our cuts and scratches; he always put "stamp-paper" on them, calling it plaster, and we knew of no other till years later. He used the same thing for his own cuts, etc., to the end of his life, and with no ill effects.

Fig. 23. Alfred R. Wallace in 1878, at age fifty-five.
From *My Life*, 2: facing p. 98.

In 1881 we moved again, this time to Godalming, where he had built a small house which he called "Nutwood Cottage." After Croydon this was a very welcome change and we all enjoyed the lovely country round. The garden as usual was the chief hobby, and Mr. J. W. Sharpe, our old friend and neighbour in those days, has written his reminiscences of this time which give a very good picture of our father. They are as follows:

About thirty-five years ago Dr. Wallace built a house upon a plot of ground adjoining that upon which our house stood. I was at that time an assistant master at Charterhouse School; and Dr. Wallace became acquainted with a few of the masters besides myself. With two or three of them he had regular weekly games of chess; for he was then and for long afterwards very fond of that game; and, I understand, possessed considerable skill at it. A considerable portion of his spare time was spent in his garden, in the management of which Mrs. Wallace, who had much knowledge and experience of gardening, very cordially assisted him. Here his characteristic energy and restlessness were conspicuously displayed. He was always designing some new feature, some alteration in a flower-bed, some special environment for a new plant; and always he was confident that the new schemes would be found to have all the perfections which the old ones lacked. From all parts of the world botanists and collectors sent him, from time to time, rare or newly discovered plants, bulbs, roots or seeds, which he, with the help of Mrs. Wallace's practical skill, would try to acclimatise, and to persuade to grow somewhere or other in his garden or conservatory. Nothing disturbed his cheerful confidence in the future, and nothing made him happier than some plan for reforming the house, the garden, the kitchen boiler, or the universe. And, truth to say, he displayed great ingenuity in all these enterprises of reformation. Although they were never in effect what they were expected to be by their ingenious author, they were often sufficiently successful; but, successful or not, he was always confident that the next would turn out to be all that he expected of it. With the same confidence he made up his mind upon many a disputable subject; but, be it said, never without a laborious examination of the necessary data, and the acquisition of much knowledge. In argument, of which intellectual exercise he was very fond, he was a formidable antagonist. His power of handling masses of details and facts, of showing their inner meaning and the principles underlying them, and of making them intelligible, was very great; and very few men of his time had it in equal measure.

But the most striking feature in his conversation was his masterly application of general principles: these he handled with extraordinary skill. In any subject with which he was familiar, he would solve, or suggest a plausible solution of, difficulty after difficulty by immediate reference to fundamental principles. This would give to his conclusions an appearance of inevitableness which usually overbore his adversary, and, even if it did not convince him, left him without any effective reply. This, too, had a good deal to do, I am disposed to conjecture, with another very noticeable characteristic of his which often came out in conversation, and that was his apparently unfailing confidence in the goodness of human nature. No man nor woman but he took to be in the main honest and truthful, and no amount of disappointment—not even losses of money and property incurred through this faith in others' virtues—had the effect of altering this mental habit of his.

His intellectual interests were very widely extended, and he once confessed to me that they were agreeably stimulated by novelty and opposition. An uphill fight in an unpopular cause, for preference a thoroughly unpopular one, or any argument in favour of a generally despised thesis, had charms for him that he could not resist. In his later years, especially, the prospect of writing a new book, great or small, upon any one of his favourite subjects always acted upon him like a tonic, as much so as did the project of building a new house and laying out a new garden. And in all this his sunny optimism and his unfailing confidence in his own powers went far towards securing him success.—J. W. S.

"Land Nationalisation" (1882), "Bad Times" (1885), and "Darwinism" (1889) were written at Godalming, also the series of lectures which he gave in America in 1886–7 and at various towns in the British Isles. He also continued to have examination papers* to correct each year—and a very strenuous time that was. Our mother used to assist him in this work, and also with the indexes of his books.

We now began to make nature collections, in which he took the keenest interest, many holidays and excursions being arranged to further these engrossing pursuits. One or two incidents occurred at "Nutwood" which have left clear impressions upon our minds. One day one of us brought home a beetle, to the great horror of the servant. Passing at the moment, he picked it up saying, "Why, it is quite a harmless little creature!" and to demonstrate its inoffensiveness he placed it on the tip of his nose, whereupon it imme-

---

* For many years he was Examiner in Physiography at South Kensington.

diately bit him and even drew blood, much to our amusement and his own astonishment. On another occasion he was sitting with a book on the lawn under the oak tree when suddenly a large creature alighted upon his shoulder. Looking round, he saw a fine specimen of the ring-tailed lemur, of whose existence in the neighbourhood he had no knowledge, though it belonged to some neighbours about a quarter of a mile away. It seemed appropriate that the animal should have selected for its attentions the one person in the district who would not be alarmed at the sudden appearance of a strange animal upon his shoulder. Needless to say, it was quite friendly.

A year or so before we left Godalming he enlarged the house and altered the garden. But his health not having been very good, causing him a good deal of trouble with his eyes, and having more or less exhausted the possibilities of the garden, he decided to leave Godalming and find a new house in a milder climate. So in 1889 he finally fixed upon a small house at Parkstone in Dorset.

Planning and constructing houses, gardens, walls, paths, rockeries, etc., were great hobbies of his, and he often spent hours making scale drawings of some new house or of alterations to an existing one, and scheming out the details of construction. At other times he would devise schemes for new rockeries or waterworks, and he would always talk them over with us and tell us of some splendid new idea he had hit upon. As Mr. Sharpe has noted, he was always very optimistic, and if a scheme did not come up to his expectations he was not discouraged, but always declared he could do it much better next time and overcome the defects. He was generally in better health and happier when some constructional work was in hand. He built three houses, "The Dell" at Grays, "Nutwood Cottage" at Godalming, and the "Old Orchard" at Broadstone. The last he actually built himself, employing the men and buying all the materials, with the assistance of a young clerk of works; but though the enterprise was a source of great pleasure, it was a constant worry. He also designed and built a concrete garden wall, with which he was very pleased, though it cost considerably more than he anticipated. He had not been at Parkstone long before he set about the planning of "alterations" with his usual enthusiasm. We were both away from home at this time, and consequently had many letters from him, of which one is given as a specimen [not included here]. His various interests are nearly always referred to in these letters, and in not a few of them his high spirits show themselves in bursts of exuberance which were very characteristic whenever a new scheme was afoot. The springs of eternal youth were forever bubbling up afresh, so that to us he never grew old. One of us remembers how, when he must have been about 80, someone said, "What a wonderful old man your father is!" This was quite a shock, for to us he was not old. . . .

~ Notes ~

Introduction ~ Biographical Sketch

1. Marchant 1916, 57.

2. People close to Wallace were well aware of his reticence to talk about himself. His first fiancée broke off their engagement because she believed that his reticence meant that he was hiding something, and even his children complained that he revealed little of himself (Wallace 1905, 1:410).

3. Mechanics' Institutes were part of a national program to educate and control the working class.

4. Robert Owen (1771–1858) was a British reformer, socialist, and author of many works, including *A New View of Society, or Essays on the Principle of the Formation of the Human Character* (1813). He organized a community at New Lanark in England according to his reformist principles (of which Wallace was a great admirer) and later at New Harmony, Indiana.

5. Alexander von Humboldt and A. G. Bonpland, *Personal Narrative of Travels* (London, 1827); William Swainson, *A Treatise on the Geography and Classification of Animals* (London, 1835); Charles Darwin, *Journal of Researches into the Geology and Natural History of the Various Countries Visited by H. M. S. Beagle* (London, 1839).

6. Robert Chambers, *Vestiges of the Natural History of Creation* (London, 1844). An enormously popular and controversial book, *Vestiges* argued that natural laws explain not only astronomical and geological phenomena, but also the historical development of all living things. It was published anonymously until Chambers's name was revealed in the twelfth edition in 1884. See Secord 2001.

7. Allen 1994; Druin and Bensaude-Vincent 1996; Larsen 1996.

8. Wallace's letters are reproduced and discussed in Brooks 1984; Bates's letters in Dickenson 1996.

9. Henry Walter Bates, *The Naturalist on the River Amazons* (London: John Murray, 1863).

10. Many of Wallace's original pencil drawings of palms and fishes from the Amazon region are reproduced in the recent book by Knapp (1999).

11. Although there is little evidence of anything but good relations for many years between Wallace and Bates, the letter of congratulation from Bates to Wallace on the publication of his 1855 "Law" paper contains the following hint of a former collaboration: "The theory I quite assent to, and, you know, was conceived by me also, but I profess that I could not have propounded it with so much force and completeness" (Marchant 1916, 53).

12. George (1991) describes the changes Wallace and others made in the place-

ment of the boundary line; Camerini (1993) provides a detailed history of Wallace's Line.

13. Edward Bellamy, *Looking Backward* (1887); *Equality* (1897).

14. Wallace 1905, 2:406. Wallace's argument was first published as "Human Selection" in the *Fortnightly Review* (1890) then as "Human Progress, Past and Future," in *Arena* (1892), and then in condensed form in his *Studies Scientific and Social* (1900).

15. Charles Dudley Warner, *Library of the World's Best Literature: Ancient and Modern* (New York: International Society, 1896), 45 vols. For other publication details of *The Malay Archipelago*, see Bastin 1986 and Smith 1991, 530.

16. Wallace received the Royal Medal (1868), the Darwin Medal (1890, for his independent origination of the origin of species by natural selection), the Copley Medal (1908), and the Order of Merit (1908) all from the Royal Society; the Gold Medal (1892) and the first Darwin-Wallace Medal from the Linnean Society of London (1908); the Founder's Medal from the Royal Geographical Society (1892); honorary doctorates from Dublin (1882) and Oxford (1889); and election to the Royal Society (1893). He was buried at a small cemetery near his home in Broadstone in December 1913 and was joined there by his widow the following year.

17. Winter 1998.

18. The bulk of Wallace's correspondence is in the manuscript collection at the British Library; see also Marchant 1916, McKinney 1972, and Raby 2001.

19. Wallace 1869, 425.

One ⚓ Wales

1. Moore 1997, 301–5.

2. The Rebecca riots peaked in the early 1840s, although there were outbreaks until the 1880s. The origin of the name is from the Old Testament, Genesis 24:60: "And they blessed Rebekah and said unto her, let the seed possess the gates of those which hate them." The rioters dressed in women's clothes to act the part of Rebecca in attacking the toll gates.

3. Wallace 1905, 1:256.

Two ⚓ The Amazon

1. William H. Edwards, *A Voyage up the River Amazon, Including a Residence at Pará* (London: John Murray, 1847).

2. Wallace 1905, 2:61.

3. Brooks (1984) describes Wallace's efforts to verify this pattern.

Three ⚓ The Malay Archipelago

1. The sixteen-year-old Charles Allen served as Wallace's assistant for a year and a half and then traveled on his own, collecting for Wallace toward the end of his stay

in the region. Allen remained in Singapore, married and had a family, and died there in about 1877.

2. These accounts were published in three separate articles in the *Annals and Magazine of Natural History* in 1856, as well as in later publications, including a lengthy account in Chapter Four of his *Malay Archipelago*. See Smith 1991 for full citations; recent discussions of Wallace's orangutan collecting are found in Van Oosterzee 1997 and in Daws and Fujita 1999.

3. Some forty-five years after Wallace's departure, the zoologist Thomas Barbour met an old Malay man who introduced himself as "Ali Wallace" (Barbour 1950, 36). That Ali took on the name of his master was not an uncommon custom for head-servants since the seventeenth century, but that he kept it for so long suggests that his relationship with Wallace was of lasting importance.

4. *The Correspondence of Charles Darwin* (Cambridge: Cambridge University Press, 1990) 6:387, 457.

5. *The Correspondence of Charles Darwin* (1991) 7:107.

6. Ibid.

7. Wallace's letter is missing; see ibid., 108.

8. In principle, Wallace's paper could have been published on its own, and by current standards this would have established his priority for the discovery of the theory of evolution by natural selection. For Wallace, this was hardly the point be-cause he knew that Darwin had been collecting evidence for a theory he had been developing for some twenty years. At the Linnean Society's fiftieth anniversary cel-ebration of the Darwin-Wallace 1858 paper, Wallace said, "I have long since come to see that no one deserves either praise or blame for the ideas that come to him, but only for the actions resulting therefrom" (Marchant 1916, 96). Although some may disagree with Wallace's ideas about what one deserves credit for, he handled "a prickly piece of controversial scientific history with impeccable tact, good nature, and sincerity" (Williams-Ellis 1966, 221).

9. Marchant 1916, 93.

Four ⚮ The World

1. Young 1985; Bowler 1986; and Van Riper 1993.

2. A. R. Wallace, "On the Varieties of Man in the Malay Archipelago," *Transac-tions of the Ethnological Society of London*, n.s., 3 (1865):196–215; printed abstracts were published in 1863 and 1864; for full publication details, see Smith 1991; "The Origin of Human Races and the Antiquity of Man Deduced from the Theory of 'Nat-ural Selection,'" *Journal of the Anthropological Society of London* 2 (1864):158–70. For more about Wallace's views on man see Fichman 1981; Bowler 1986; and Rich-ards 1987.

3. Gould 1980.

4. His book *On Miracles and Modern Spiritualism*, published in 1875, was revised and reprinted several times. During the 1860s and 1870s, he held seances at his home, testified in court on behalf of mediums, and published a variety of letters and articles defending the claims of spiritualists. See Milner 1996 and Raby 2001.

5. Wallace's 1856 paper on the habits of the orangutan, as mentioned in the introduction to Chapter Three, shows that he had been wrestling with this for more than a decade. Interesting discussions of Wallace's so-called conversion are found in Richards 1987; Smith 1991, 1999; and Fichman 2001.

6. Wallace 1905, 2:116.

7. Ibid, 160.

◄ Books by Alfred Russel Wallace ►

Only the first editions of Wallace's books are listed here. For complete bibliographic information on these books and on all of Wallace's publications, see Charles Smith's *Alfred Russel Wallace*, 1991.

1853   *Palm Trees of the Amazon and Their Uses.* London: John Van Voorst.

   A Narrative of Travels on the Amazon and Rio Negro, with an Account of the Native Tribes, and Observations on the Climate, Geology, and Natural History of the Amazon Valley. London: Reeve and Co.

1869   *The Malay Archipelago: The Land of the Orang-utan and the Bird of Paradise, a Narrative of Travel with Studies of Man and Nature.* London: Macmillan and Co.; New York: Harper and Brothers.

1870   *Contributions to the Theory of Natural Selection: A Series of Essays.* London: Macmillan and Co.

1875   *On Miracles and Modern Spiritualism: Three Essays.* London: James Burns.

1876   *The Geographical Distribution of Animals: With a Study of the Relations of Living and Extinct Faunas as Elucidating the Past Changes of the Earth's Surface.* 2 volumes. London: Macmillan and Co.; New York: Harper and Brothers.

1878   *Tropical Nature, and Other Essays.* London: Macmillan and Co.

1879   *Australasia* (edited and extended by Alfred R. Wallace, with Ethnological Appendix by A. H. Keane). Stanford's Compendium of Geography and Travel. London: Edward Stanford.

1880   *Island Life: Or, the Phenomena and Causes of Insular Faunas and Floras, Including a Revision and Attempted Solution of the Problem of Geological Climates.* London: Macmillan and Co.

1882   *Land Nationalisation: Its Necessity and Its Aims: Being a Comparison of the System of Landlord and Tenant with That of Occupying Ownership in Their Influence on the Well-being of the People.* London: Trübner and Co.

Books by Alfred Russel Wallace

1885   *Bad Times: An Essay on the Present Depression of Trade, Tracing It to Its Sources in Enormous Foreign Loans, Excessive War Expenditure, the Increase of Speculation and of Millionaires, and the Depopulation of the Rural Districts; with Suggested Remedies.* London: Macmillan and Co.

1889   *Darwinism: An Exposition of the Theory of Natural Selection with Some of Its Applications.* London: Macmillan and Co.

1891   *Natural Selection and Tropical Nature: Essays on Descriptive and Theoretical Biology.* London: Macmillan and Co.

1898   *The Wonderful Century: Its Successes and Its Failures.* London: Swan Sonnenschein and Co.; New York: Dodd, Mead, and Co.

1900   *Studies Scientific and Social.* 2 volumes. London: Macmillan and Co.; New York: The Macmillan Co.

1903   *Man's Place in the Universe: A Study of the Results of Scientific Research in Relation to the Unity or Plurality of Worlds.* London: Chapman and Hall; New York: McClure, Phillips, and Co.

1905   *My Life: A Record of Events and Opinions.* 2 volumes. London: Chapman and Hall; New York: Dodd, Mead, and Co.

1907   *Is Mars Habitable? A Critical Examination of Professor Percival Lowell's Book "Mars and Its Canals," with an Alternative Explanation.* London: Macmillan and Co.; New York: The Macmillan Co.

1908   *Notes of a Botanist on the Amazon and Andes* (by Richard Spruce, edited and condensed by Alfred Russel Wallace). 2 volumes. London: Macmillan and Co.

1910   *The World of Life: A Manifestation of Creative Power, Directive Mind, and Ultimate Purpose.* London: Chapman and Hall.

1913   *Social Environment and Moral Progress.* London: Cassell and Co.

         The Revolt of Democracy. London: Cassell and Co.

⤙ *Bibliography* ⤚

The single best source for additional readings on Wallace can be found on the Web at the Alfred Russel Wallace Page.

Allen, David. 1994. *The Naturalist in Britain: A Social History,* 2nd ed. Princeton: Princeton University Press.

Barbour, Thomas. 1950. *Naturalist at Large.* London: Scientific Book Club.

Bastin, John. 1986. Introduction to Wallace's *Malay Archipelago,* vii–xxvii. New York: Oxford University Press.

Beddall, Barbara B. 1968. Wallace, Darwin, and the theory of natural selection. *Journal of the History of Biology* 1:262–323.

Beddall, Barbara B. 1972. Wallace, Darwin, and Edward Blyth: Further notes on the development of evolutionary theory. *Journal of the History of Biology* 5:153–58.

Beddall, Barbara B., ed. 1969. *Wallace and Bates in the Tropics.* London: Macmillan.

Bowler, Peter. 1986. *Theories of Human Evolution: A Century of Debate, 1844–1944.* Baltimore: Johns Hopkins University Press.

Brackman, Arnold C. 1980. *A Delicate Arrangement: The Strange Case of Charles Darwin and Alfred Russel Wallace.* New York: Times Books.

Brooks, John L. 1984. *Just Before the Origin: Alfred Russel Wallace's Theory of Evolution.* New York: Columbia University Press.

Browne, Janet. 1983. *The Secular Ark: Studies in the History of Biogeography.* New Haven: Yale University Press.

Camerini, Jane R. 1993. Evolution, biogeography, and maps: An early history of Wallace's Line. *Isis* 84:700–27.

Camerini, Jane R. 1996. Wallace in the field. *Osiris* 11:44–65.

Camerini, Jane R. 1997. Remains of the day: Early Victorians in the field. In *Victorian Science in Context,* edited by B. Lightman, 354–77. Chicago: University of Chicago Press.

Clements, Harry. 1983. *Alfred Russel Wallace: Biologist and Social Reformer.* London: Hutchison and Co.

Daws, Gavan, and Fujita, Marty. 1999. *Archipelago: The Islands of Indonesia, from the Nineteenth-Century Discoveries of Alfred Russel Wallace to the Fate of Forests and Reefs in the Twenty-first Century.* Berkeley: University of California Press and Nature Conservancy.

Desmond, Adrian. 1989. *The Politics of Evolution: Morphology, Medicine, and Reform in Radical London.* Chicago: University of Chicago Press.

Dickenson, John. 1996. Getting on in his rambles in South America: The published correspondence of H. W. Bates in the Amazon 1848–59. *Archives of Natural History* 23(2):201–8.

Bibliography

Druin, Jean-Marc, and Bensaude-Vincent, Bernadette. 1996. Nature for the people. In *Cultures of Natural History*, edited by N. Jardine, J. A. Secord, and E. C. Spary, 408–25. Cambridge: Cambridge University Press.

Fichman, Martin. 1981. *Alfred Russel Wallace.* Boston: Twayne Publishers.

Fichman, Martin. 2001. Science in theistic contexts: a case study of Alfred Russel Wallace. *Osiris* 16:227–50.

George, Wilma. 1964. *Biologist Philosopher: A Study of the Life and Writings of Alfred Russel Wallace.* London: Abelard-Schuman.

George, Wilma. 1980. Alfred Wallace, the gentle trader: collecting in Amazonia and the Malay Archipelago, 1848–1862. *Journal of the Society for the Bibliography of Natural History* 9:515–25.

George, Wilma. 1991. Wallace and his line. In *Wallace's Line and Plate Tectonics*, edited by T. C. Whitmore, 3–8. Oxford: Oxford University Press.

Gould, Stephen Jay. 1980. Wallace's fatal flaw. *Natural History* 89(1):26–40.

Knapp, Sandra. 1999. *Footsteps in the Forest: Alfred Russel Wallace in the Amazon.* London: Natural History Museum.

Larsen, Anne. 1996. Equipment for the field. In *Cultures of Natural History*, edited by N. Jardine, J. A. Secord, and E. C. Spary, 358–77. Cambridge: Cambridge University Press.

Marchant, James, ed. 1916. *Alfred Russel Wallace: Letters and Reminiscences.* New York: Harper and Brothers.

McKinney, H. Lewis. 1972. *Wallace and Natural Selection.* New Haven: Yale University Press.

Milner, Richard. 1996. Charles Darwin and Associates, Ghostbusters. *Scientific American* 275:72–77.

Moore, James. 1997. Wallace's Malthusian moment: the common context revisited. In *Victorian Science in Context*, edited by B. Lightman, 290–311. Chicago: University of Chicago Press.

Parker, Percy L., ed. 1912. *Character and Life.* London: Williams and Northgate.

Quammen, David. 1996. *The Song of the Dodo.* New York: Scribner.

Raby, Peter. 2001. *Alfred Russel Wallace: A Life.* London: Chatto & Windus; Princeton: Princeton University Press.

Richards, Robert J. 1987. *Darwin and the Emergence of Evolutionary Theories of Mind and Behavior.* Chicago: University of Chicago Press.

Seaward, M. R. D., and Fitzgerald, S. M. D., eds. 1996. *Richard Spruce (1817–1893): Botanist and Explorer.* London: Royal Botanical Gardens, Kew.

Secord, James A. 2001. *Victorian Sensation: The Extraordinary Publication, Reception, and Secret Authorship of Vestiges of the Natural History of Creation.* Chicago: University of Chicago Press.

Smith, Charles H. 1992, 1999. Alfred Russel Wallace on Spiritualism, Man, and Evolution: An Analytical Essay. Torrington, Conn. (pamphlet).

Smith, Charles H., ed. 1991. *Alfred Russel Wallace: An Anthology of His Shorter Writings.* Oxford: Oxford University Press.

Van Oosterzee, Penny. 1997. *Where Worlds Collide: The Wallace Line.* Ithaca, N.Y.: Cornell University Press.

Bibliography

Van Riper, A. Bowdoin. 1993. *Men among the Mammoths: Victorian Science and the Discovery of Human Prehistory.* Chicago: University of Chicago Press.

Williams-Ellis, Amabel. 1966. *Darwin's Moon: A Biography of Alfred Russel Wallace.* London: Blackie.

Winter, Alison. 1998. *Mesmerized: Powers of Mind in Victorian Britain,* Chicago: University of Chicago Press.

Young, Robert M. 1985. *Darwin's Metaphor: Nature's Place in Victorian Culture.* Cambridge: Cambridge University Press.

~*~ Index ~*~

Numbers in *italics* denote illustrations.

Index

About the Editor

Jane R. Camerini is an independent scholar in the history of science and a faculty associate at the University of Wisconsin, Madison, where she received her Ph.D. in biocartography. She has written numerous journal articles, reviews, and book chapters on the history of evolutionary natural history, focusing on field work, biogeography, mapping, and the work of Alfred Russel Wallace and other nineteenth-century naturalists. She and her husband have a home in Madison and a farm in rural Wisconsin.